PLAIN SOLDIERING

MILITARY BOOKS BY THE SAME AUTHOR

Before the Echoes Die Away
(The story of a Territorial Gunner Regiment)

Gunners at Larkhill
(A history of the Royal School of Artillery)

PLAIN SOLDIERING

A history of the Armed Forces on Salisbury Plain

N.D.G. JAMES

SALISBURY
THE HOBNOB PRESS

First published in 1987 by
The Hobnob Press
5 Bridge Street, Salisbury SP1 2ND England

British Library Cataloguing in Publication Data
James, Noel David Glaves
Plain Soldiering
Military History

ISBN 0 946418 03 9

Printed and bound by
Butler & Tanner Ltd, Frome and London

Contents

Foreword

by LIEUTENANT GENERAL SIR MICHAEL GRAY KCB OBE
General Officer Commanding South East District

It is interesting to muse on the contribution that Salisbury Plain has made to the fighting skills of the British Army since it returned from fighting the Boers in South Africa in 1902. The area was used well before that for military manoeuvres but what is certain, is, that during this century there must be very few British soldiers who have not trained there at one time or another. Commanders, at all levels, have applied their tactical knowledge and teased the skill of their soldiers over those windswept grassy slopes for nigh on a hundred years.

In 1982 when Salisbury Plain came under my aegis as the District Commander of the South West, I discovered that a history of British Army activity on and around the Plain did not exist. Many an experience had been recorded in journals and regimental histories over the years, but none of this had been collated. I therefore asked each of the Garrisons, bounding the Training Area, to start some research work and begin collecting together archive material. Later, I invited the author to have a look at what we had found and if he considered it viable, to help us write up the record. The result is this book. You will discover, as you read that he has himself dug very deeply and researched meticulously. Regimental histories, the Army List, old maps, local press reports, the Land Agent's records, Ministry of Defence Minutes and, of course, the memories of those who live thereabouts, have all been tapped and what has emerged is a fascinating journal of our military past. It reflects the ebb and flow of military budget and the effect of this and technology on organization and tactics. The narrative highlights the pressures that are imposed on training facilities as armies expand rapidly, are reinforced by Allies and need time to practice the Art of War and, when the war is over, how the peacetime political pressure to acquire farmland and to build houses is in conflict with the needs of Defence. Starkly brought out are the difficulties faced by the Army when changes in the technology of weaponry and cross country mobility increase the importance of training space. An area that once catered for the manoeuvres of over 50,000 troops is now hardly capable of holding a single Brigade of 5,000 men.

The author has taken great care to look at each of the barracks, at where and how they came to be built and named and, perhaps more importantly, how they are laid out and have developed over the years. His narrative records in great detail stories about the formations and units who have lived and trained there. He tells of the massive Reviews of horses, men and bands at the end of the 19th Century and of the later well known Tidworth Tattoo. Little anecdotes and quotations about people bring to life this amazing treatise about the Army and the Plain as it was and, indeed, now is.

Few will also know that the first Army aerodrome was constructed on Salisbury Plain in 1910 and that the first Military Air Trials took place from Larkhill in 1912. These were the links in the chain of events which indirectly led to the formation of the Royal Flying Corps, later to become the Royal Air Force.

What is also interesting is the remarkable way in which the military ownership of this land has preserved so much of the natural flora and fauna. These chalky uplands and valleys have been protected from modern farming methods and from insecticides with the result that plants, flowers, birds and insects, which have disappeared from most other areas, now flourish. It is a conservationist's paradise.

This book is a most readable record which brings to life the story of our Army's most famous training area, Salisbury Plain. It is also a valuable reference document from which to cull lecture material for units who arrive in the training camps and barracks and are about to embark on new adventures on this much used and indeed much loved part of England.

OCTOBER 1986

Preface

When Rudyard Kipling that great champion of the British Army, wrote

> Scent of smoke in the evening
> Smell of rain in the night
> The hours, the days and the seasons
> Order their souls aright

he might well have been thinking of Salisbury Plain where for generations, the Army has learnt and practised its battlecraft and tactics. Indeed there can be few soldiers – serving or retired – who have not taken part in exercises on the Plain or fired their weapons on one of its many ranges whilst pilots of countless aircraft, have been familiar with the skies above it.

This book is a history of the Plain from the Services' point of view but it should be remembered that today, times are continually changing and technology is constantly advancing. However, in the last analysis, it is the soldier who wins battles whether it be at Paardeberg, Alamein or Goose Green: *Plain Soldiering* is essentially about such men and their times.

N.D.G.J.
Blakemore House,
Kersbrook,
Budleigh Salterton,
Devon.
November 1986.

Acknowledgements

Unfortunately space does not allow me to mention by name, all those who have helped me during the course of writing this book but I am especially indebted to the following.

The Andover Local History Society; the Librarian Andover Public Library; Maj. R.D. Baker, 16/5L; Sqn. Ldr. K.Baldock., RAF; Maj. P.J. Ball; the Marquess of Bath; Mr. C.Bayliffe; Mrs. Berkery; Mr. C.J.J. Berry; Mr. E.C. Booker; Mr. P.B. Boyden; Mr. F.M. Broadbent; Maj. D.B.S. Brotherton; Mr. J.F. Budzynski; Mr. Bernard Bull; Maj. W.G.J. Byott, RAMC; Mr. L.Campbell; Mr. F.E. Card; Capt. D.A. Chance; Mr. H.Chandler and the staff of Reprographics, Royal School of Artillery; Mr. K.E. Chant; Maj. E.Chapman; Maj. N.L. Chesher; Brig. C.Childs, CBE; Mrs. J.A. Churchill; Maj. J.L.F. Clarke, MBE, TD; Maj. D.K.R. Clifton-Moore; Mr. Paul Cole; Mrs. H.E.P. Colley; Mrs. P.Colman, Librarian, Wiltshire Archaeological and Natural History Society; Mr. B.A. Cordery; Lt.-Col. J.B. Cowan; The Rev. J.W. Croft; Maj. J.R. Cross AAC; Mrs. K.P. Cummins; Lt. R.S. Davis, RAOC; the Editor, *Debrett's Peerage*; Col. P.R.R. de Burgh, OBE, Librarian, Badley Library, Royal School of Artillery; Capt. K.E. De Witt, US Army; Mrs. N.Dibben; Capt. D.J. Eadie, 16/5L; Mrs. V.M. Edwards, Regional Secretary NAAFI; Mr. A.Eyres; Lt.-Col. A.A. Fairrie; Lt.-Col. D.N. Faithful-Davies, RAMC; Maj. D.J.H. Farquaharson, 16/5L; Brig. J.P. Ferry; Col. J.S. Fielding, MBE; Maj. J.Firbank, Archivist, The Salvation Army; Lt.-Col. D.Fisher, MBE; the Curator Fleet Air Arm Museum; Brig. B.N.L. Fletcher, CBE; Mr. David Fletcher; Maj. M.J. Fogwell; the Rev. C.R. Wolsey Gilbert, Secretary, Methodist Church Forces Board; Mr. J.S. Gledhill; Mr. I.J. Gollings, National Secretary, The Returned Services League of Australia; Lt.-Col. M.H. Goodhart; Mr. G.H. Griffin; Mr. E.Hall; Mrs. K.Harrison; Mr. T.H.J. Heffernan; Col. P.W. Herring, OBE; Col. F.M. Hill, CBE; the Rev. J.J. Holliman, SCF; Capt. P.Hughes, RAOC; Col. R.D. Hughes; Lt.-Col. A.Hunt; Mr. J.W. Hunt, Librarian Royal Military Academy Sandhurst; the Librarian, The India Office Library and Records; Lt.-Col. J.M. Jago, OBE; Maj. R.J. Jeffreys, MBE; the Kennet District Council; Mr. D.Kyte;

the Managing Director Kiwi Products (UK) Ltd., Maj. M.W. Knight, RAOC; Miss H.M. Knox, General Secretary, Sandes Soldiers' and Airmens' Homes; Maj. D.S. Langley, MC; Capt. S.J. Leach, R SIGNALS; Mr. B.M. Little, Librarian Salisbury Library; The Ludgershall History Society; Col. A.M. McCaffery, BEM; Lt.-Col. J.S. McLaren, OBE; Mrs. A. McMenemy; the Chief Librarian, Ministry of Defence, Whitehall Library; Mr. J. Moloney; Maj. W. Murison, RE; Maj. F. Myatt, MC; Lt.-Col. S.M.H. Nash, OBE; the Director, the National Army Museum; Lt.-Col. N.J. Newman, RAEC; The New Zealand High Commission; Flt.-Lt. J. O'Donnell, RAF; The Rev. J. O'Hara, CF (RC); The Ordnance Survey; Mr. B.A. Painter, MBE; Mr. M.W. Parry; Lt.-Col. R.S.G. Perry, DSO; Maj. J.H. Peters, MBE, Curator of The Duke of Edinburgh's Royal Regiment's Museum; Mr. R. Peters; Mrs. J. Philpots; Mr. K. Pickernell, BEM; Brig. J.R.I. Platt, DSO, OBE; Mr. G.R. Poulton; Mr. A.E. Price; the Staff of Property Services Agency; Lt.-Col. F.P. Riordan; Lt.-Col. C.C.G. Ross; the Director, Royal Air Force Museum; Mr. Alan Russell; Maj. P.L. Rutherford, QARANC; the Director of Infantry; Mr. M.A. Shallcross; Col. T.C. Sherry, OBE; Mr. T. Sinclair; Capt. D.M. Skelly, US Army; Col. R.J. Skinner; Messrs. Daniel Smith, Chartered Surveyors; the Rev. M.J.B. Spencer, CF; WO2(SQMS) J.A. Stanistreet; Mrs. S.M. Stirling, Librarian, Devon & Exeter Institution; Maj. A. Stockley, SASC; Mr. F. Stone, OBE; Mr. Michael Stratton; Maj. R.B. Sullivan-Tailyour; Mr. D.J. Tempero; Mrs. P. Thacker; Maj. S.G. Thornborough, RAEC; Maj. J.J. Thornton, TD; Mr. L.E. Trevett; Mr. N.E. Trowell, Hon. Curator NAAFI Historical Collection; Mr. P.H. Vickers, Librarian, Prince Consort's Army Library; Mr. John Waight; Mr. G. Ward; The Estate Manager, Whitbread Wessex; Maj. M.E. Whitelock; Mr. G.E. Wickens; Maj. M. Wicksteed; Flt.-Lt. S.M. Wiles, WRAF; Lt.-Col. D.J. Wilson, MBE, AAC; the Chief Education Officer, Wiltshire County Council; the County Archivist, Wiltshire County Council; Maj. W.J.A. Withers; the Editor, *Who's Who*; Lt.-Col. R.G. Woodhouse; Lt.-Col. W.F. Woodhouse; Capt. M. Woods, RAOC.

I also wish to thank the following for permission to quote or use information from the publications stated or to use photographs or maps, details of which are given in the lists of references or illustrations.

The Editor, *After the Battle* – 'Incidents at Imber' by W.G. Ramsey.
W.H. Allen, *The Silver Road* by Mary Harrison.
The Editor, *Andover Advertiser*, various reports.
The Director, Australian War Memorial, Canberra, *Anzac to Amiens* and *The Australian Imperial Force in France 1916*, both by C.E.W. Bean.
Blandford Press Ltd., *The British Infantry 1660–1945* by Frederick Myatt.
Frank Cass & Co. Ltd., *Modern English Biography.*
H.C. Casserley – four photographs.
The Director-General, Commonwealth War Graves Commission – Records of Cemeteries and Churchyards.
Constable & Co. Ltd., *Smith-Dorrien* by Collin Ballard.
The Defence Land Agent, one photograph.
Andre Deutsch, *Russell's Dispatches from the Crimea 1854–1856* edited by Nicolas Bentley.
T.L. Fuller & Sons, seven photographs.

The Editor, *Gunner* – 'Observing Fire from Balloons' by Lt.-Col. J.B. Cowan and a photograph of HRH The Prince of Wales.
Hamish Hamilton Ltd., *The Moon's A Balloon* by David Niven.
The Imperial War Museum, ten photographs.
The Longman Group UK Ltd., *The Science of War 1891–1903* by Col. G.F.R. Henderson edited by Col. N. Malcolm.
Norman Longmate, *The G.I's – The Americans in Britain 1942–1943*.
Lutterworth Press, *Reluctant Missionary* by Edith Buxton.
Gertrude McCracken, *Looking Back on Seventy Years of Sutton Veny.*
Methuen & Co. Ltd., *Roundabout Wiltshire* by A.G. Bradley.
The Director, Directorate of History, National Defence Headquarters, Ottawa, Canada, *Fighting Newfoundlanders* by Col. G.W.L. Nicholson.
Penguin Books Ltd., *Queen Victoria's Little Wars* by Byron Farwell.
The Curator, Queen Alexandra's Royal Army Nursing Corps Museum, *Proceedings and Reports.*
The Colonel, 16th/5th The Queen's Royal Lancers, two photographs.
Regimental Home Headquarters, 4th/7th Royal Dragoon Guards, three photographs.
Regimental Headquarters, Royal Tank Regiment, extracts from *The Tank Corps Journal*, the *Royal Tank Corps Journal* and *The Tank.*
Regimental Secretary, Queen's Own Highlanders (Seaforth and Camerons), one photograph.
The Librarian, The RE Corps Library, *The Royal Engineers' Journal*, vol. LXVI, 1952 which contained articles by Brig. P.W.L. Broke-Smith that later appeared in book form under the title of *The History of Early British Military Aeronautics.*
The Editor, *Salisbury Journal*, various reports.
The Librarian, The Tank Museum, two photographs.
John W.R. Taylor, *C.F.S. – Birthplace of Air Power.*
Thorsons Publishing Group Ltd., *Action Stations. 5* by Chris Ashworth.
Headquarters, United Kingdom Land Forces, two photographs.
The Director, *The Warminster Journal*, various reports.
Whitcoulls Ltd., Auckland, New Zealand, *The War Effort of New Zealand* by Lt. H.T.B. Drew.
Wild Swan Publications Ltd., and David Bartholomew, *Midland & South Western Junction Railway.*
The Editor, *Wiltshire Gazette and Herald*, various reports.

In conclusion, I wish to acknowledge the help and advice that I have received on many occasions from my son, Maj. T.D.R. James, RA.

PUBLISHER'S ACKNOWLEDGEMENTS

The enthusiasm and generosity of the following organizations and individuals have enabled this book to be published:

DOWNING RUDMAN & BENT (SALISBURY PLAIN) LTD
Syrencot House, Figheldean, Salisbury, Wiltshire
Building and public works contractors in Wiltshire for 150 years

CARTON INDUSTRIES (1944) LTD
Bulford Road, Durrington, Salisbury, Wiltshire
Protective packaging specialists

Churchfields Dairies Ltd, Salisbury, Wiltshire
Michael Snell Patisserie Tea Rooms, Salisbury, Wiltshire
Tesco Stores Ltd, Andover, Hants

This book was conceived by Lieutenant General Sir Michael Gray, KCB, OBE, when he was GOC South West District and, in addition to writing the Foreword, he has helped enormously in its eventful birth. In this latter role he was ably assisted, initially by Colonel Tom Sherry of the Royal Army Education Corps in Bulford and subsequently by Colonel Ian Mackay, Commander Education at South East District, Aldershot.

Mr Charles Shea-Simonds and Dr John Chandler have made important contributions. Miss Alison Borthwick is responsible for the excellent cartography.

Illustrations

Maps

CHAPTER ONE

The Great Plain

The extensive area of chalk downland which lies in the south of Wiltshire has, for many centuries, been known as Salisbury Plain or, more locally, as 'The Plain'. Its boundaries do not appear to have been defined very precisely and some differences of opinion have existed as to its exact extent. In *Kelly's Directory of Wiltshire* which was published in 1898, the Plain is said to reach 'from the north of Salisbury for about sixteen miles or a little south of Pewsey and from east to west for some twenty miles from Ludgershall on the east to Westbury on the west'. In fact the latter distance is nearly twenty-five miles. On the other hand Ella Noyes, a Wiltshire writer, considered that

> Salisbury Plain properly only comprises the South Wiltshire downs, and a few miles of Hampshire, where the western edge of the county runs out into Wiltshire. Its limits may be defined as from Red Horn Hill above Urchfont to Salisbury north and south, and from Mere to Ludgershall east and west. It is a great wedge-shaped chalk plateau, bounded at its widest end, on the north, by the Vale of Pewsey and the wooded country of the Bristol Avon and its tributaries, on the west of the broken district once covered by the old Forest of Selwood, on the east by the valley of the Bourne and by the Nadder on the south-west.[1]

The important difference between these two descriptions is the fact that Ella Noyes considered that the Plain extended as far as Mere. She then states that the western boundary is 'the broken district once covered by the old forest of Selwood'. In fact Selwood Forest lay astride the county boundary being partly in Wiltshire and partly in Somerset. The eastern boundary of the Wiltshire portion ran southwards from Bulkington near Devizes, to the east of Erlestoke Park then through Imber and Knook to East Knoyle. The eastern boundary of Selwood in Somerset followed the county boundary from Rode near Trowbridge southwards to the Frome road (B3092) and then deviously close to Maiden Bradley, Stourhead and Zeals, two miles west of Mere.[2] It would seem probable that in her definition of Salisbury Plain, Ella Noyes was referring to the Forest of Selwood in Somerset and not Selwood in Wiltshire and this would agree with her reference to Mere. In view of this together with her opinion that the River Nadder formed the south-west boundary, it is clear that she considered that the triangle of land lying between Warminster, Mere and Wilton to be part of the Plain. Disregarding the fact that in 1984 this area is intensively farmed and that it includes two large areas of woodland – Grovely and Great Ridge – it is not difficult to visualize it as part of Salisbury Plain.

Today the Plain is generally considered to be the area which lies within the following bounds. On the north, the road from Westbury to Rushall continuing by the road through Upavon to Ludgershall. The eastern boundary is formed by the road from Ludgershall to North Tidworth (A3026) and then by the road through Cholderton to a point one mile south of Winterbourne Earls. From here the southern and south-western boundary follows a line to Old Sarum; then across the Salisbury–Devizes road until it meets the Salisbury–Warminster road (A36) near South Newton which it follows to Warminster and so back to Westbury. It is with reference to this area that the name Salisbury Plain is used, in the course of this book.

However, when John Speed published his map of Wiltshire in 1610, he showed two areas which were inscribed 'Salesburye Playne'.[3] The larger of these is in effect, the Plain as it is known today while the centre of the smaller area was some five miles south of Salisbury. Its approximate boundary ran from Knighton Wood near the Blandford–Salisbury road (A354), northwards to the River Ebble which it followed to its junction with the Avon. From there the boundary continued southwards along the river to Wick near Downton where it turned westward to rejoin Knighton Wood. When the first edition of Andrews and Dury's *Map of Wiltshire* was published in 1773 the smaller area was clearly marked as 'Salisbury Plain' as was also the case in the second edition of 1810.[4] Thomas Davis in his account of farming in Wiltshire in 1794[5] includes a map showing both of these areas inscribed as 'Salisbury Plain' as Cary does on his 'new map of Wiltshire' of 1801.[6] However, when the first Ordnance Survey map (to the scale of one inch to one mile) was published in 1811, it did not name the smaller area and there does not appear to have been any further reference to this 'little Salisbury Plain'.

The area of the Plain which lies within the present day boundaries that have already been described, amounts to approximately 227·50 square miles which is the equivalent of 145,600 acres or 58,820 hectares. Of this, some 91,000 acres (36,825 hectares) belonged to the Ministry of Defence in 1984, which corresponded to ·625 of the whole.[7] For the most part Salisbury Plain consists of chalk downland which geologically is part of the Upper Cretaceous System. With the exception of Sidbury and Beacon Hills, the highest features extend from Warminster Down

692ft. (211m.) above sea level in two directions. Eastwards following the northern boundary through Tinhead Hill 738ft. (225m.) and Urchfont Hill 712ft. (217m.) to Casterley Camp which stands above the Avon Valley, and south-eastwards, through Battlesbury Hill 682ft. (208m.) Codford Castle and Yarnbury Castle 548ft. (167m.). Between these two extended features of high ground, the land falls gradually to a height of about 262ft. (80m.) besides the River Avon. Beyond the river the ground rises again to 460ft. (140m.) or so in the centre, and 558ft. (170m.) on the northern boundary, reaching a height of 735ft. (224m.) at Sidbury Hill, the highest point on the Plain. To the south, Beacon Hill 670ft. (204m.) forms the last major feature as the land falls away to heights of about 328ft. (100m.) and correspondingly lower in the valleys.

Five rivers are related to the Plain. The Avon which flows through Upavon, Netheravon and Amesbury is joined at Bulford by the Nine Mile River. The Bourne rises to the north of Ludgershall and after passing through Tidworth continues to Shipton Bellinger and Cholderton, and along the southern boundary, the Wylye is joined at Stapleford by the Till which rises near Orcheston.

As regards archaeology there can be few areas of greater interest than Salisbury Plain where Stonehenge, innumerable barrows and much evidence of the beginnings of civilization are to be found. However these are matters which have been covered by many authorities and lie beyond the scope of this book.

It is probable that this country has seen greater changes during the course of the present century than at any other period of its history. Many have been brought about by wars or the threats of wars and of these, Salisbury Plain has experienced its full share. Today it is difficult to visualize what it was like before the Services entered upon the scene and the Plain formed an immense, silent and remote expanse of countryside. A rather brief and prosaic description appeared in *Kelly's Directory* for 1899 to the effect that

> The Salisbury downs or plains form a high bleak ground with scanty grass, and unpeopled except in the valleys of the rivers where the towns and hamlets lie close together.[8]

Fortunately a rather more illuminating account has been provided by A.G. Bradley and this gives some idea of what the Plain was like on an August afternoon, eighty years ago.

> Another four miles of chalk road rising and falling over the open down; more far-reaching vistas of this strange land to the east, and to the west, and to the south; nothing else to be seen but the long billows of green patched here and there upon their sides, or in their troughs, with golden harvest fields, and tufted with the clumps of fir or beech, which in the far distance show merely as black specks upon the waste. There are no roadside fences here to break the view, and the low grassy banks which fringe the

> highway are rank by now with the tall stiff stems and matted undergrowth of cocksfoot, ryegrass or fescues, and decked in patches here and there with the modest hues of scabious and wild parsley, but mostly powdered white in this August weather by the dust of travelling sheep or the occasional motor. No grouse crow, nor do curlews call, amid these chalk solitudes; nor yet is there anywhere the faintest note of tumbling water. Indeed, the characteristic quality of the Wiltshire downland is, as often as not, its overpowering silence.[9]

Until the Army became established on the Plain, agriculture was the principal occupation and consequently the local population was largely dependent on it, either directly or indirectly. The wheelwright, the blacksmith, the saddler and the corn merchant all looked to the farmer for much of their livelihood, quite apart from those who were actually employed on the land. At the same time shopkeepers in the neighbouring towns, as well as in the villages, did much of their business with the farming community. The majority of those who lived in the villages were also connected with agriculture in one way or another and an outstanding description of village life at the end of the last century has been provided by Flora Thompson in her classic *Lark Rise to Candleford.*

The late Mr. Eric Sargent was only six weeks old when, in 1896, his father became tenant of Enford Farm near Netheravon which belonged to Sir Edmund Antrobus of Amesbury Abbey. He was subsequently granted the joint tenancy of the farm with his brother in 1927, after the death of his father, and continued to occupy it until he died in 1978. During his later years he wrote an account of life at Enford Farm in the early days and the following is taken from his description of the farmhouse.

> The open fireplace (in the Kitchen) was used for roasting joints, poultry, etc., in Dutch ovens, the spits of which revolved mechanically by winding up a spring. Meat and poultry cooked in this manner and kept basted, had a flavour which one does not get in the ovens and cooking appliances today.
>
> All the water for the baths was carried from the copper in the back kitchen to the bedrooms and of course had to be carried down again...
>
> Lighting was either by oil lamps or candles and the work entailed in connection with this, cleaning and trimming, etc., was considerable.[10]

Salisbury Plain was the home of the Wiltshire Horned Sheep which were also known as the Western or Horned Crock. In his report on farming in Wiltshire which was published in 1794, Thomas Davis who was Steward to the Marquess of Bath at Longleat, recorded that

> The kind of sheep which are chiefly kept in South Wiltshire is what has been long known in Smithfield Market by the name of the Wiltshire Horned Sheep.[11]

Both sexes were horned but their fleece was light and they carried little wool with none at all on their undersides due, it was said to the dry warm chalky soil. Despite the fact that in 1849 the breed was considered to be 'now nearly extinct'[12] it was still in existence in 1984.

The successful management of a flock depended on the shepherd and many were completely dedicated to their work, as at Enford Farm.

> The shepherd, John Holmes, who was well over seventy, walked daily from the village before pitching his fold and taking out his sheep, in every sort of weather. He never took a holiday, working seven days a week, and often informed me how many years he had worked without a break. And this, despite the very heavy snow storms which were much more common than at present. He was so regular that one could put one's time right by his comings and goings. He was a fine man with patriarchal features and must have been bred from some good stock. He knew all his sheep although to the outsider they are all so much alike, and he seemed able to keep the identity and history of each one in his memory. Sometimes for a number of days he would meet no one with whom he could converse between the time he left his home in the morning and his return in the evening. Under these circumstances one can appreciate the mutual love and understanding between the shepherd and his dog.[13]

The traditional system of farming which had been practised on the Plain for many generations, was known as 'sheep and corn'. This entailed the sheep being grazed on the downland during the summer and being folded on the more sheltered arable land in the winter, usually from October to March. A variation of this procedure was adopted on some farms where the flock would remain on the downs during the day, returning to the folding area in the late afternoon. Folding or penning consisted of erecting hurdles or sheep netting so as to enclose part of a field of roots, generally turnips, an area of one square chain (22yds × 22yds) being allowed for two hundred sheep per day. Early every morning a new area would be enclosed and the sheep moved into it, and in this way they were provided with a fresh supply of turnips, the ground received a concentrated application of manure from the sheep and the land could be ploughed as soon as the hurdles were moved forward. After ploughing it was sown to wheat and so completed the cycle of sheep and corn.

Until the end of the seventeenth century, Salisbury Plain consisted almost entirely of grassland on which immense flocks of sheep grazed, but in 1724 Daniel Defoe noted that changes were taking place.

> The vast flocks of sheep which one everywhere sees upon these downs and the great number of those flocks, is a sight truly worth observation; 'tis ordinary for these flocks to contain from three to five thousand in a flock and several farmers have two or three flocks.
>
> But 'tis more remarkable still, how a great part of these downs comes by a new method of husbandry, to be not only made arable which they never were in former days but to bear excellent wheat, and great crops too tho' ... never known to our ancestors to be capable of any such thing ... For by folding the sheep upon the plow'd lands, those lands ... are made fruitful

and bear very good wheat as well as rye and barley.[14]

The conversion to arable continued intermittently until the time of the Napoleonic Wars but with the return of peace some of the areas reverted to pasture. However the old practice of improving poor land by 'burnbaking' or 'burn-beaking' which had been carried out for many decades, still continued. This consisted of removing the turf with a mattock and burning it, the ashes being spread over the ground. In this way the land was prepared for an arable crop and fertilized at the same time. Evidence of burn-baking is to be found in several place names such as Netheravon Bake, Rowden's Bake Barn and Rollestone Bake Farm.

In the valleys, water meadows formed an important part of the farming economy, their construction and maintenance requiring the expert knowledge and life long experience of those who looked after them who were very appropriately known as 'drowners'. These meadows which were only suitable for grazing by sheep since cattle would trample in the water channels, could produce a heavy yield of grass early in the season. A notable grassland area was at Orcheston known as the Wiltshire Long Grass Meads, which although only two and a half acres in extent produced 'immense' crops of exceptional quality.[15]

John Aubrey in his description of the downs which was written between 1656 and 1691, remarked that although 'the turfe is of sweet short grasse, good for the sheep and delightfull to the eye', there was 'not a tree or rarely a bush to shelter one from a shower'.[16] He further expressed the opinion that

> long time ago Salisbury Plaines might have woods of them, but that they cut them down as an incumbrance to the ground which would turn to better profit by pasture and arable.[17]

Over the next one hundred and fifty years or so, the situation does not appear to have changed and William Cobbett writing from East Everleigh in August 1826 declared that

> this is real down-country. Here you see miles and miles square without a tree or hedge or bush. It is a country of greensward.[18]

Cobbett's observation agrees with the evidence provided by the Ordnance Survey map which was published in 1817, to a scale of one inch to one mile. This shows only eight small woods or groups of plantations on the land which now forms the Ministry of Defence training area, all of which are named on the map. To the west of the road from Shrewton to West Lavington (A360) there were three: Wadman's Coppice, a mile north-west of Imber; Fir Clump on Clay Pit Hill, a mile to the south of Chitterne church and Gilbert's Copse a mile and three quarters due east of Fir Clump. All these are shown on the 1:25,000 Ordnance Survey maps of 1958 (sheets ST94 and SU04) although Wadman's

1. SALISBURY PLAIN BEFORE THE WAR OFFICE ACQUISITIONS

From an Ordnance Survey map published in 1900
Scale: 4 inches to 1 mile

Coppice has alone retained its name and only a small part of Gilbert's Copse is shown near Bushes Farm.

Between the Shrewton–West Lavington road and the Amesbury–Upavon road (A345) three areas are marked on the 1817 map: New Coppice, a mile to the east of St. Joan à Gore farmhouse; a small square enclosure named 'Firs' about one third of a mile south-west of Ell Barrow and Talbot Plantations, a group of six scattered enclosures a mile to the north-west of Casterley Camp. The first two are shown on the 1:25,000 map (sheet SU05) but only New Coppice has retained its name. To the east of the Amesbury–Upavon road two woodland features are marked and named on the map of 1817: Fir Plantations, a group of six small enclosures a mile to the east of Fifield, and Hills Coppice on Beacon Hill which is named on the 1:25,000 map (sheet SU24) as Hills Copse.

In 1901 a second edition of the Ordnance Survey map drawn to the scale of six inches to one mile, was published and this shows that a large number of woods, plantations and shelter belts had been planted since 1817. These comprise many of the older woodland areas such as Shrewton Folly and Dunch Hill Plantation and it is clear that these areas were planted by previous owners before the War Department acquired the land.

In conclusion, another glimpse of the Plain at the beginning of the century, is provided by the map on page 00 which was published by the Ordnance Survey in 1900. None of the later army centres are shown on it and the only 'camps' which are marked are those of archaeological interest. At this date four railways encompassed the Plain three of which are still in public use and details of these will be found in Chapter 13. The road from Salisbury to Devizes by way of The Bustard and Ell Barrow was in regular use and Imber was an active community.

CHAPTER TWO

The New Order

During the latter half of the nineteenth century fundamental changes began to take place in the organization and administration of the British Army. A number of these arose from the public concern which had been expressed regarding the blunders and inefficiency that had come to light during the Crimean War. In addition to making alterations in the composition and operation of the War Office, steps were taken to train officers in staff duties by the establishment of the Staff College in 1855. At the same time, the Small Arms Factory at Enfield was enlarged and brought up to date while improvements in the design of soldiers' uniforms were accompanied by the formation of an Army Clothing Department.

Although in the course of the next few years the impetus of these changes diminished to some extent, it was restored in 1868 when Edward Cardwell (afterwards Viscount Cardwell) was appointed as Secretary of State for War by Gladstone. Under his able and energetic direction, a programme of far-reaching reforms was put into effect. These included the termination of the system under which commissions could be purchased; the introduction of short-service enlistments of six years with the colours and six with the reserve; the reorganization of infantry regiments by counties; the linked battalion system and the abolition of flogging in peacetime. Improvements in pay and living conditions were later made by H.C.E. Childers who was Secretary of State for War from 1880 to 1882. By the end of the century further progress had been made in re-organization, the Army Service Corps (now the Royal Corps of Transport) being formed in 1880, the Army Pay Corps (now the Royal Army Pay Corps) in 1893 and the Royal Army Medical Corps in 1898. While these changes were in themselves important steps forward, they also reflected a new line of thought on the need to create an efficient and professional army. To attain this, adequate training was essential and after 1850 a new approach to this problem began to emerge.

In 1851 at the instigation of Lord Hardinge who had succeeded the Duke of Wellington as Commander-in-Chief of the British Army, a survey was made of the countryside near Aldershot and arising out of this, an experimental camp was established in the following year, at Chobham Common, some six miles north-east of Camberley.[1] The purpose of this was to assess the efficiency of the troops concerned and their methods of training and this was so successful that a suggestion was put forward for continuing it as a permanent camp. During the summer of 1853 the largest exercises which had so far been held, took place in the Aldershot area when some twenty-five thousand troops were involved.[2] These proved to be as valuable as those held in the previous year and in January 1854 it was announced that agreement had been reached for the purchase of Aldershot Heath and work commenced on the erection of permanent barracks in the following September.[3] Later, during the 1860s, a large area of land was acquired for training purposes between Aldershot and Fleet which included the Long Valley and Laffan's Plain.[4]

The next development was the establishment of the School of Musketry at Hythe which officially came into being on 1st April 1854 under its first commandant, Lieut.-Col. C.C. Hay. The main object of the School was to train instructors in musketry who were classified as Captain Instructors, Lieutenant Instructors, Sergeant Instructors and Instructors. After completing an eight weeks course in a satisfactory manner, they returned to their units but remained on the establishment of the School and under the direct control of the Commandant.[5]

Admiral of the Fleet, Lord Fisher, GCB, OM, GCVO attended a course at Hythe as a naval lieutenant and has left the following account of it.

> In 1861, when I was sent to the Hythe School of Musketry as a young lieutenant, I found myself in a small squad of officers. My right-hand man was a General and my left-hand man a full Colonel. The Colonel spent his time drawing pictures of the General. (The Colonel was a wonderful artist). The General was splendid. He was a magnificent man with a voice like a bull and his sole object was Mutiny! He hated General Hay who was in command of the Hythe School of Musketry. He hated him with a contemptuous disdain. In those days we commenced firing at a target only a few hundred yards off. The General never hit the target once! The Colonel made a beautiful picture of him addressing the Parade and General Hay: 'Gentlemen! My unalterable conviction is that the bayonet is the true weapon of the British soldier!' The beauty of the situation was that the General had been sent to Hythe to qualify as Inspector-General of Musketry.
>
> After some weeks of careful drill (without firing a shot) we had to snap caps (that was to get our nerves all right, I suppose!); the Sergeant Instructor walked along the front of the squad and counted ten copper caps into each outstretched hand. At that critical moment General Hay appeared on the

> Parade. This gave the General his chance! With his bull-like voice he asked General Hay if it was believable after these weeks of incessant application that we were going (each of us) to be entrusted with copper caps!
>
> When we were examined viva voce we each had to stand up to answer a question (like the little boys at a Sunday School). The General was asked to explain the lock of the latest type of British rifle. He got up and stated that as he was neither Maskelyne and Cooke nor the Davenport Brothers (who were the greatest conjurors of that time) he couldn't do it. Certainly we had some appalling questions. One that I had was 'What do you pour the water into the barrel of the rifle with when you are cleaning it?' Both my answers were wrong. I said 'With a tin pannikin or the palm of the hand'. The right answer was 'With care!' Another question in the written examination was 'What occurred about this time?' Only one paragraph in the text-book had those words in it 'About this time there occurred, etc!'
>
> All the same I had a lovely time there; the British Army was very kind to me and I loved it. The best shot in the British Army at that date was a confirmed drunkard who trembled like a leaf but when he got his eye on the target he was a bit of marble and 'bull's eyes' every time! So, as the Scripture says, 'Never judge by appearance!'[6]

In 1919 the School was re-named the Small Arms School and further information about it will be found in later chapters.

The absence of suitable training areas and firing ranges placed brigades and batteries of the Royal Artillery at a particular disadvantage and although the School of Gunnery was formed at Shoeburyness on 1st April 1859, it provided little scope for fire and movement as it was sited on the coast and faced the sea. Consequently the use of Woolwich Common for exercises and Plumstead Marshes for live firing continued for some years and it was not until 1875 that the first Gunner practice camp was set up at Okehampton.[7] Two years later this was followed by 'Camps for Artillery Instruction' at Aldershot, Hay and Romney.[8] In 1882 another permanent practice camp was established at Lydd but by this date temporary practice camps were apparently being set up on any local sites which appeared to be suitable for the purpose.[9]

There were several reasons why it was now a matter of some urgency that the Army should be provided with adequate training areas, if its future efficiency was to be assured. As the British Empire continued to extend its boundaries, increasing demands were made on the armed forces to assist in the maintenance of law and order, to deal with organized resistance to British rule and to meet any form of aggression. Farwell has compiled a list of hostilities which took place during the sixty-three years of Queen Victoria's reign. This records thirty-six wars, eighteen campaigns, ninety-eight expeditions and thirteen operations, making a total of one hundred and sixty-five but it does not include numerous occurrences which he has described as sieges, insurrections, disturbances, rebellions and uprisings.[10]

At home there was the need to provide for the defence of the Country and there were justifiable reasons for bearing this in mind. During and after the Crimean War relations with the French had seriously deteriorated while the German victory over the French in the Franco-Prussian War of 1870, created further apprehension although this stemmed from a different source. The revival of the Volunteer Movement in 1859 and the passing of the Reserve Forces Act 1882,[11] were evidence of the Government's concern. Under Cardwell's system of linked battalions, one was stationed at home while the other served overseas and for which the home-based battalion trained reinforcements. Once again the emphasis was on training and effective training required the necessary facilities.

Although unit training could be carried out on comparatively restricted sites, large formation exercises which were usually referred to as 'manoeuvres', required an area extending over many square miles. By 1897 manoeuvres had been held in several localities including Aldershot in 1871 and 1892, Salisbury Plain in 1872 and Dartmoor in 1873. Before any troops could use the area which had been selected for the exercises a considerable amount of preparatory work had to be done. Apart from obtaining permission from the owners to use the land, arrangements had to be made for the erection of any temporary buildings, the use of private roads, the provision of adequate water supplies and so on. A sufficient supply of water was a matter of the greatest importance as the Army was entirely dependent on horses for transport. In order to regulate these matters the Military Manoeuvres Act was passed in 1897 which was described simply as 'An Act to facilitate Army Manoeuvres'.[12] Although it was amended in 1911[13], the greater part of it remained in force until it was finally repealed by the Manoeuvres Act 1958.[14]

The first manoeuvres to be held after the passing of the Act of 1897 took place on and adjacent to Salisbury Plain from 1st to 8th September 1898. These were described as 'the greatest military assembly since 1871', 50,301 troops taking part as compared with 32,769 at Aldershot in 1871.[15] In 1904 there was a complete change of location, when the manoeuvres were held in Essex in order to practice defence against an invading seaborne force – a forerunner of 1940. However during 1907 and 1908 they returned once again to Salisbury Plain. In 1907 a small booklet which provided an entertaining account of the manoeuvres of that year, entitled *Tha Military Manoovers*, was written in Wiltshire dialect by Edward Slow of Wilton under the pen-name of 'Measter Benjamin Sloper'. He described the strategy of the manoeuvres as follows.

> Tha plan as vur as I can zee is this. Tha Invadin Army (Blue), ten thousand strong, under Zir Ian Hamilton, is zappoused to have landed at Malberer, an is gwain ta march right across Zalsbury Plaain ta try an Konker tha Defenden Army (Rid), we about tha seam number a troops under Zir Frederick Stopvird, who is hurrien up from down Bournemouth way to stop

> tha Invaders an drave em back inta tha sae agaen.[16]

An illustration of the cover is shown on plate 4.

Extensive open spaces were also required from time to time, where reviews of large numbers of troops could be held and these sometimes took place at the end of the manoeuvres. A review was in effect, a march past on a massive scale in which several thousand troops participated and at which a very senior officer took the salute . The designation 'royal review' was sometimes applied when the monarch or another member of the Royal Family was the reviewing officer. One of the first major reviews was held on 12th September 1872 at Bulford below Beacon Hill, the slopes of which provided a natural grandstand.

> On Thursday the march past took place and this brought to an end the autumn manoeuvres of 1872 ... The scene was Beacon Hill and a better ground for such a purpose could not be found or desired. Early in the morning the streets of the city were alive with people starting on foot for the scene of the march past and by eight o'clock many heavily laden vehicles had left the city on the way to Beacon Hill.[17]

It is interesting to read that many of those who lived in Salisbury were quite prepared to walk to Beacon Hill, a distance of about twenty miles there and back; something which few people would consider doing today.

> At noon, a royal salute ... announced the arrival of the Prince of Wales who was received by the Duke of Cambridge and a very numerous staff of general officers including the foreign military representatives ... About one o'clock the march past commenced ... the last regiment came to the saluting point at twenty minutes to three o'clock, the passing of the 30,000 men thus occupying one hour and forty minutes.[18]

The march past was followed by a trot past of the mounted troops after which 'a grand parade of the whole army took place' and 'thus ended the grandest spectacle that was ever witnessed from the far-reaching ken of Beacon Hill'.

The next review was held on 8th September 1898 but on this occasion the site had been moved from Beacon Hill to Boscombe Down. Some fifty-thousand troops were assembled at Porton Down, three miles to the south-east of Boscombe Down and then marched to the reviewing area. It took two and a half hours for all the troops to pass the saluting base. The new location was not popular with the general public since it did not provide such a good view of the proceedings.

> It appears that the final choice of ground was influenced by the opinion that though 30,000 troops had plenty of room at Beacon Hill for the evolutions of 1872, the movements of the 50,000 men of 1898 would have been cramped there.[19]

In spite of this, the occasion aroused great interest and was watched by thousands of spectators; it was described afterwards as 'the Review of the Century'. A description of the day written in verse and entitled *Mrs. Brown at the Review*, was published and sold for one penny.

> We've bin up top o' Boscombe Down
> About three mile fra' Amesbury town,
> An' there we seed a splendid sight,
> A hundred thousan' soldiers quite
> A marching by so gallantly.[20]

It would appear however that Mrs. Brown had rather overestimated the number of troops taking part.

Towards the end of the nineteenth century it became clear that action would have to be taken to provide a sufficiently large area of land on which military training could take place at any time of the year and over which the army had complete control. One of the leading military writers of the period, Col. G.F.R. Henderson, CB claimed that 'the cry throughout the army from 1870 onwards' was 'Give us the men to command and the ground where we may train ourselves and them!'[21]

After the defeat of the British by the Boers at Majuba Hill in 1881 a number of improvements had taken place in training.

> The annual allowance of ammunition for target practice had been largely increased. Drill in close order had been relegated to its proper place; the time given to spectacular movements had been greatly reduced, and the barrack-square, whenever men and space were available, was deserted for the open country.[22]

In spite of this, there was still an urgent need in 1895 for greater opportunities for field and formation training which in turn depended on better facilities, that is to say, larger training areas. 'Duties about barracks had still the first claim upon the men and field exercises when compared with those of Continental armies were limited in both scope and duration'.[23]

One of the first steps to be taken in the provision of new training areas was the passing of the Military Lands Act of 1892. This gave the Secretary of State for War, powers to purchase and lease land for military purposes, to make byelaws in respect of such land and to deal with other related matters.[24] This was followed by three amending Acts, namely the Military Land Acts of 1897, 1900 and 1903.

After the Act of 1892 had become law, considerable thought was given to the acquisition of a new training area and Salisbury Plain was recognized as being particularly suitable for the following reasons. Much of it consisted of a large expanse of mainly unenclosed land having a light soil over chalk which dried out quickly and over which horses and horse transport could move without difficulty. On the whole it was not considered to be an area of any great agricultural value and as much of it was only used as sheep-walks, it commanded an acceptably low price thereby reducing the cost of acquisition. From a military point of view it was well placed, London being eighty miles distant, Aldershot fifty miles and the great seaport of Southampton only thirty.

Although rumours had been circulating for some time, it was not until 29th January 1897 that the Government's intentions were made known. Speaking in the House of Commons on that day, the Under Secretary of State for War (the Rt. Hon. W. St. John Brodrick) asked for the authorization of a loan 'for urgent military purposes' of which four hundred and fifty-thousand pounds were 'to be devoted to the acquisition of a large tract of land for manoeuvring purposes on Salisbury Plain'.[25] Reporting on this the *Salisbury and Winchester Journal* stated that confirmation had been received

> from Mr. Powell Williams, Financial Secretary of the War Office, which stated that the Government had purchased 40,000 acres of land on Salisbury Plain for the purpose of establishing a permanent military training ground there. We understand that Messrs. Rawlence and Squarey have been acting for the War Office in this matter for the past two months. The large tract of land that has been acquired, some sixty square miles in extent, is stated to be situated to the north of Amesbury and around Figeldean and has recently been inspected by Sir Evelyn Wood, Adjutant-General, Mr. Powell Williams and Mr. St. John Brodrick, the Under Secretary for War.[26]

A week later the *Journal* contained further information regarding the extent of the land that was being acquired.

> We have learned on what we believe to be good authority that the Government are buying a considerable amount of land east of the Avon including part of Sir J. Kelk's Tedworth property; and as they also appear to be acquiring a large tract of downland to the west of the river, the inference is that the more valuable land which is being purchased in the Avon Valley near Figeldean ... is to serve to connect the eastern and western portions of the manoeuvring ground.[27]

As regards the use to which this land was to be put the *Journal* continued

> From the statement of Mr. Brodrick in the House of Commons on Friday night, the 29th ult..... it would appear that the primary purpose of the new training ground will be to serve as a manoeuvring ground for cavalry. Infantry, however, will also be trained, a rifle range will be established and it is hoped in addition to have artillery practice on the Plain.[28]

Today the Plain presents a rather different picture.

The purchase of so large an area of land by the Government raised a number of important questions and these were set out in a minute issued by the Secretary of State for War (the Marquess of Lansdowne) on 29th April 1897. In this he asked

> What additional land do we require in the Avon Valley in order to bring the two sections within striking distance (if that is the proper expression)?
>
> How are we going to deal with the population within the area to be acquired?

How do we propose to manage and administer so large a landed estate as that for which we shall be responsible?

What are we going to do with the house at Tedworth, and its surroundings?[29]

To consider these questions and other related matters, Lord Lansdowne appointed a small committee, known as the War Office Salisbury Plain Committee of which the following were members

The Parliamentary Under-Secretary of State: the Rt. Hon. W.St. John Brodrick, MP as President.
The Financial Secretary: J. Powell Williams, MP.
The Adjutant General: Gen. Sir Redvers H. Buller, VC, GCB, GCMG.
The Quartermaster-General: Gen. Sir H. Evelyn Wood, VC, GCB, GCMG.
The Inspector-General of Fortifications: Maj.-Gen. Sir Robert Grant, KCB.

with Maj. Graham Thomson, RE as Secretary.[30]

At the same time the Secretary of State drew the attention of the Committee to several matters which he considered should be borne in mind when reaching any decisions. Although the land was being acquired for military purposes, he felt that 'a wholesale disturbance of the agricultural population' should be avoided. He saw no reason why most landowners should not retain their houses with the land in the immediate vicinity, as well as their fishing and possibly their shooting. As regards the farmers, although the downland would be required for military purposes, he hoped that they would be able to continue cultivating the better land around their homesteads and if so, there should be little reduction in the employment of farm workers. If any land became vacant, consideration should be given to letting parts of it as small holdings to farm workers and village tradespeople. Finally in view of the excellent condition of Tedworth House, its buildings and gardens, he hoped that a use could be found for it but if not, that a suitable purchaser would be forthcoming so as to ensure that it did not fall into disrepair.[31] All these points indicated the genuine concern of the Government that the acquisition of the land should cause as little disturbance as possible, subject always to the main consideration, namely, its use for military purposes.

The first meeting of the Committee was held on 11th May 1897 and it continued to operate for the next five years during which time thirty-three meetings were convened, the last taking place on 25th April 1902. Although for the whole of its existence the constitution of the Committee was based on the appointments which the five original members held, those who served on the Committee in these appointments changed from time to time due to retirements, transfers and so on. In April 1902 the Committee was composed of entirely different individuals to those who had served on it in May 1897.

The public reaction to the announcement that the Government intended to purchase large areas of the Plain was for the most part favourable, although there were some exceptions. These were among the landowners whose land was to be acquired and who were consequently more affected by these new developments than other members of the public. When reporting on the matter, the *Salisbury Journal* stated that

> while local opinion is generally favourable to the scheme, some of the resident gentry of the district more immediately concerned, are understood to regard with much disfavour the prospects of a military inroad. Sir E. Antrobus is stated to have a great dislike of the Government proposals, and another gentleman residing nearer the probable centre, is believed to be very hostile to them.[32]

In the next issue of the *Journal*, the gentleman who was 'believed to be very hostile' was named as G. Knowles of Syrencote House, Figheldean.[33]

Speaking at a meeting at Figheldean on 7th February, Lord Folkestone, the local Member of Parliament said that

> as he understood it, the farms would still remain in the hands of the present tenants at reduced rentals ... He did not see why the interests of the working classes should be affected by the camp and in regard to the shopkeepers, he thought it would improve their position and be good for the trade of their district.[34]

As is usual on such occasions, a certain amount of correspondence was generated in the local press and in one letter which appeared over the initials 'H.M.' (later identified as Sir Henry Malet of Wilbury Park, Newton Tony) the writer expressed the following opinion.

> The greater part of the land in question has long been practically unsaleable on any terms. A few residents no doubt will have to be expropriated, but there can be no fear but that they will be liberally compensated. As regards the district at large nothing but good can result. There will be a revival of movement and prosperity all around it. The markets will be stimulated, small cultivators of fruit and vegetables and poultry owners will benefit immensely, and the decaying villages and farms on Salisbury Plain will ere long, participate in the general improvement ... I for one rejoice that Salisbury Plain is likely to realise its manifest destiny as the training ground of the British Army.[35]

From the beginning of August 1897 to the end of December 1899 the purchases of forty-five properties were completed and these varied in size from a few acres to several thousand. The total area acquired amounted to 41,021 acres (16,572 hectares) the cost of which was £513,178. The first land to be bought was Grove Farm of some 364 acres situated on Ball Down in the parish of West Lavington which belonged to a body known as Legge's Trustees. There were no buildings on the land and the price was agreed at £3,647, that is ten pounds an acre, the purchase being completed on 3rd August 1897. This land now forms a part of West Down artillery range. By the end of 1897 an area of some 13,661 acres had been acquired and this included several substantial properties, the most important being the Tedworth Estate of 6,618 acres which was purchased on 29th September. Belonging to Sir John William Kelk, Bt. It included Tedworth House, 13 farms, 8 farmhouses, 107 cottages and the Ram Hotel in North Tidworth and was bought for £95,000. Another substantial block of land situated at Orcheston and Tilshead, extending to 2,181 acres was purchased from Miss Mills three weeks later.[36]

The four largest properties acquired in 1898 were Sir Michael Hicks-Beach's Netheravon Estate of 7,813 acres which was bought for £93,411; Lord Normanton's property of 2,973 acres at Rushall and Enford purchased for £31,825; land at Bulford belonging to J. Ledger Hill amounting to 1,917 acres and the Syrencote Estate of 1,275 acres owned by G. Knowles. During 1899 two more substantial areas of land were added: the Brigmerston Estate which belonged to F.S. Rendall and covered 2,205 acres and land at Enford the property of Sir Edmund Antrobus, Bt. amounting to 2,384 acres. Of the forty-five purchases which were made before September 1899, it was necessary to use powers of compulsory purchase in the case of eight owners amongst whom were the Ecclesiastical Commissioners, Sir Michael Hicks-Beach, King's College Cambridge, G. Knowles and Winchester College.[37]

On 29th June 1899 *The Devizes and Wiltshire Gazette* published 'a map of that portion of the county of Wiltshire which is affected either directly or indirectly by the Government purchases of land for military purposes'.[38] A copy of this is shown on page 00. After the end of December 1899 the number of properties acquired decreased rapidly and by 31st March 1902 the area purchased had increased by only 908 acres, bringing the estate to a total of 41,929 acres.[39] A month later the final meeting of the Salisbury Plain Committee was held on 25th April 1902.

The rapid expansion of the War Department's estate during the first five years was followed by a period which proved to be something of a hiatus, as only 1,500 acres were bought between April 1902 and March 1910. However in the course of the next ten years the net acquisitions, that is the area purchased less any land sold, rose to nearly 8,000 acres of which 5,480 acres were bought before the end of 1912.[40] It is probable that this renewed activity was partly due to the deterioration of the situation in Europe which led to the outbreak of war in 1914.

On 27th September 1927 a significant step was taken in the expansion of the War Department's estate when the first land lying to the west of the Shrewton–West Lavington road (A360) was purchased. This was an area of 935 acres in the parishes of Imber and West Lavington and was the forerunner of the numerous acquisitions which were ultimately to provide land for the School of Infantry at Warminster and for the Imber ranges. Between 1920 and 1930 the total area which was purchased amounted to some 14,850 acres most of which lay to the

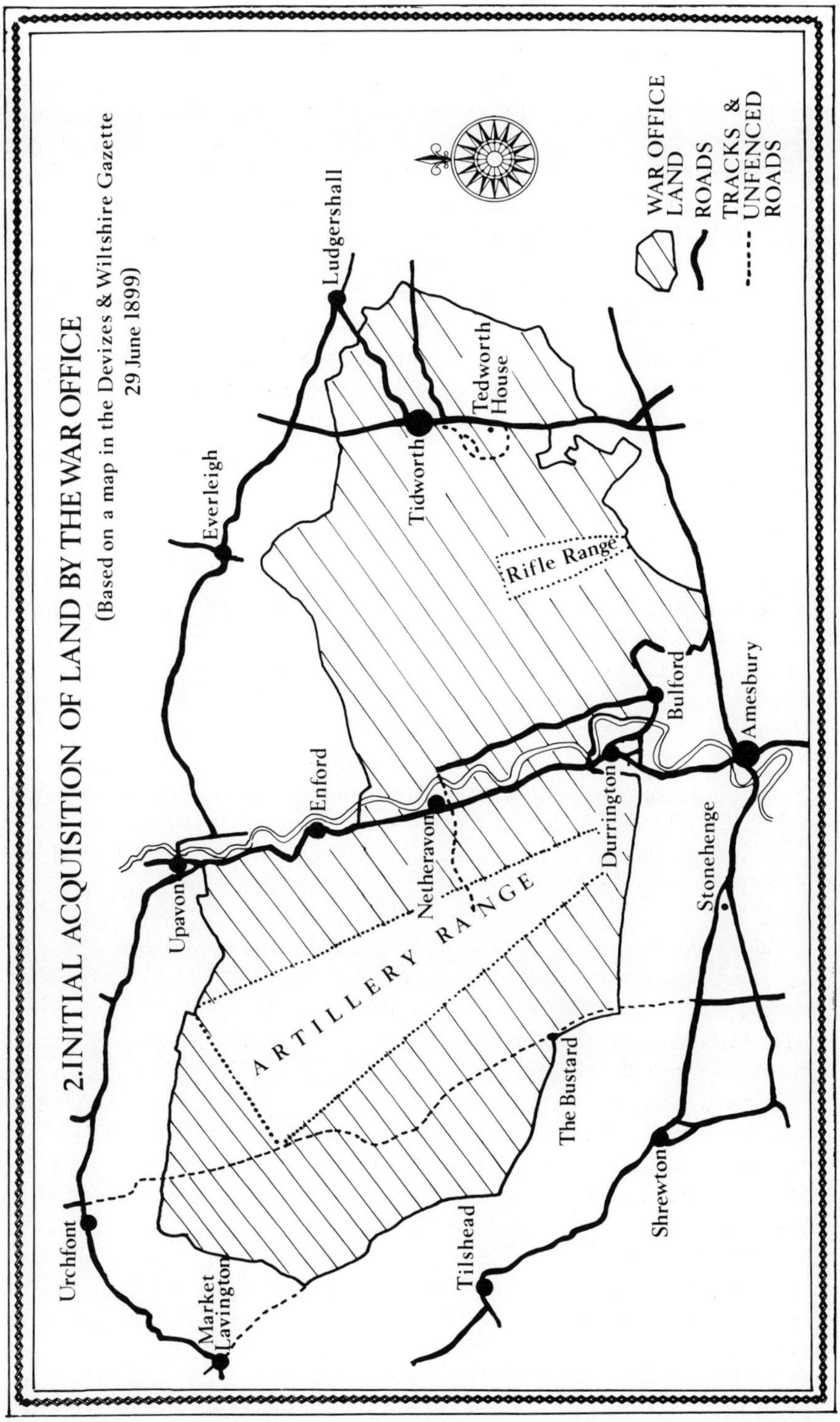
2.INITIAL ACQUISITION OF LAND BY THE WAR OFFICE
(Based on a map in the Devizes & Wiltshire Gazette
29 June 1899)
WAR OFFICE LAND
ROADS
TRACKS & UNFENCED ROADS
Ludgershall
Everleigh
Tidworth
Tedworth House
Rifle Range
Enford
Upavon
Netheravon
Bulford
Amesbury
Durrington
ARTILLERY RANGE
Stonehenge
The Bustard
Shrewton
Tilshead
Urchfont
Market Lavington

west of the Tilshead road. The three largest areas which were acquired comprised 3,611 acres purchased from Lord Heytesbury in December 1929; Manor Farm, Chitterne, 1,661 acres, bought from C.H. Collens in November 1927 and 1,165 acres acquired from the Marquess of Bath in March 1929.[41]

During and after the Second World War, purchases of land declined to such an extent that over the forty-three years from 1st April 1940 to 31st March 1983 the net increase in area amounted to only 1,537 acres. Of this total the largest single purchase was Everleigh Manor (1,115 acres) which was bought in November 1954.[42] On 31st March 1983 the area belonging to the Ministry of Defence on Salisbury Plain amounted to some 91,709 acres (37,105 hectares) and a summary of the areas acquired between 1897 and 1983 will be found in Appendix I.

CHAPTER THREE

Laying the Foundations

During the first two years which followed the appointment of the War Office Salisbury Plain Committee, properties were being acquired at the rate of nearly two a month and with such rapid expansion the problems of management soon became apparent. In addition to transacting the normal day-to-day matters of business, the underlying need was to create a well run estate out of the numerous independent areas of land that were being purchased and which varied from a few fields to properties of several thousand acres.

Tenants on the larger estates usually occupied their farms under a tenancy agreement which had been drawn up in accordance with the owner's wishes and which would consequently vary from one estate to another. However it was not uncommon at this period for some farms to be let by word of mouth instead of under a written agreement and this could make the position even more difficult. Another matter which involved a great deal of time and money was the management and repair of houses and cottages and on a large estate this frequently required the services of an experienced building staff. Decisions would also have to be made in respect of the woodlands and steps taken to deal with any shooting or fishing belonging to the estate. There were in addition, a number of less obvious aspects of land ownership which concerned the village schools, almshouses and churches on the estate. The Salisbury Plain Committee experienced all these problems during the five years of its existence. It is fortunate that the work and decisions of the Committee have been recorded in the minutes of its proceedings and these have proved to be an invaluable source of information.[1]

One of the first steps to be taken by the Committee was to appoint Messrs. Rawlence & Squarey of 8, Rollestone Street, Salisbury 'as agents for the management of the Salisbury Plain property'.[2] This well known firm of land agents had already been acting for the War Department in purchasing a number of farms and the partners who were to be chiefly concerned with the new estate were E.P. Squarey and N. Squarey. This arrangement continued until 1st January 1902 when the management was taken over by a whole time resident agent.[3]

Another urgent matter with which the Committee was concerned was the future of Tedworth House. Although its purchase from Sir John William Kelk was not completed until 29th September 1897 the Chairman reported at a meeting on 24th May of that year that 'the military authorities had decided to recommend the sale of the house and part of the land'. Consquently the Committee agreed to sell the house and contents, 2,200 acres of land, the Ram Inn and the outlying cottages by auction.[4] However, when it was put up for sale on 28th July it failed to find a purchaser and it was accordingly decided to sell it privately.[5] On 23rd September, largely on account of the views put forward by the Adjutant-General (Gen. Sir Redvers Buller, VC, GCB, GCMG) this decision was revoked and it was agreed that the property should not be sold and that a tenant should be found for Tedworth House.[6] However by 20th January 1898 all efforts to let the house had failed and at a meeting held on that day the Financial Secretary, J. Powell Williams, MP, expressed the opinion that if the house was retained 'in about twenty years it would be a most valuable property'.[7] This marked the beginning of a change in policy and in the following October 'It was unanimously decided that the house and grounds should not, under present circumstances, be sold'.[8] At the next meeting in December 1898, a new approach to the problem was considered and approved.

> The Chairman brought forward a proposal that Tedworth House should be handed over to the newly appointed Commanding Royal Engineer, Salisbury Plain, Colonel Barklie, to provide quarters for himself and staff, and accommodation for the drawing office, etc., so that the preparation of plans for the new barracks may be commenced without delay.
>
> After further discussion, it was decided that Colonel Barklie should be allowed to enter into possession of the house...[9]

Other than agreeing, in August 1901, to extend this arrangement for 'a further period of (say) one year' no other action was taken on Tedworth House before the Committee was dissolved in April of the following year.[10]

In order to improve the conditions for training it was decided that all arable land on or adjoining the training area should be sown to permanent pasture and at a meeting held on 4th August 1899 it was agreed.

> That all land outside the approved limits of the Avon Valley cultivation zone should be laid down permanently to grass, as soon as possible...

Certain limited exceptions were made to this decision which included gardens, allotments and small areas of land adjoining farm premises.[11] The minutes of this meeting also mentions 'land in Schedule III of the agreement forms (i.e. permanent pasture of which the feed only is let)'. This was the first occasion that any reference had been made to the three categories into which the War Department's land (other than that which was in regular use for training) had been divided, namely, Schedule I, II and III lands.

Schedule I comprised land let to farm tenants under an agricultural tenancy agreement, over which training rights were reserved subject to compensation being paid to the tenant for any damage that might occur. Schedule II was land which was arable when it was purchased and which the War Department directed should be laid down to grass, the Army providing the seed and the tenant agreeing to maintain the area as permanent grassland. Land which was primarily used for training purposes although it was possible to farm it on a limited scale, was placed under Schedule III and was only let under licence. The Army reserved the right to use this land whenever they wished and no compensation was paid in respect of any damage which was caused. Schedule III land was frequently let to those who were already tenants of Schedule I land. After the conversion of the original arable areas to grassland had been completed, the classification 'Schedule II land' became obsolete and was discontinued but Schedule I and Schedule III categories were retained and were still in use in 1986. In that year Schedule I lands covered an area of some 21,000 acres and Schedule III land, 42,000.

At the end of the last century Salisbury Plain supported a very large rabbit population which was not diminished to any great extent until myxomatosis occurred in 1953. If the War Department was unaware of the problem when the land was bought, it soon came to the notice of the Committee. The question of getting rid of the rabbits was first raised at a meeting held in June 1897 when it was decided that the Agents should 'arrange for operations being commenced as early as feasible'.[12] The reason for the Committee's concern was the danger which rabbit holes created on land used by cavalry and mounted troops,[13] and some idea of the problem can be gained from the fact that in 1897, some 14,000 rabbits had been killed on the Tedworth estate alone.[14] When an area of land which now forms part of West Down artillery range, was bought in 1897, a rabbit warren of some 300 acres (120 hectares) surrounded by a wire netting fence was found near The Bustard Inn. Steps were subsequently taken to remove it.[15] In November 1897 it was decided that the innumerable rabbit holes should be filled in by hand but six months later nothing had been done. This was apparently on account of the cost as it was estimated that it would take twenty men one month to complete the work.[16] Eventually it was agreed to pass this problem on to the unfortunate farm tenants who 'should be told that they must kill the rabbits down this year and fill in the holes'. After this a warrener was to be employed to catch any rabbits that were left in the open.[17]

Another matter which created a number of problems was the shooting over different parts of the estate and attempts to let it were not always successful. The Tedworth shoot was a source of continual difficulties and at one period was let from week to week, while for a short time it was let to the Aldershot Game Association and later to the Tedworth Game Association the members of which were officers from Aldershot and Tidworth.

Under Messrs. Rawlence & Squarey's guidance the Committee adopted a sound policy in respect of the estate woodlands. When in November 1897 attention was drawn to the need to cut the underwood, not only was this approved but it was also agreed to employ a second woodman to help in the work.[18] A few months later the question of thinning some of the woodlands at Tedworth was raised.

> Mr. Squarey pointed out that all that was required was the ordinary thinning requisite in forestry, and that cutting down of timber was not contemplated.[19]

Approval was given to this proposal and it was later agreed that the larch and Scots pine poles produced by these thinnings, should not be sold but supplied free of charge for the use of the War Department at Tidworth.[20] In the autumn of 1898 the Committee decided that 'the usual cutting of underwood' should be carried out and that the stock of young trees in the nursery at Tedworth should not be sold but used for planting on the estate.[21]

In the course of buying numerous farms and estates the War Department had acquired a large number of cottages and it was felt that these should be suitably labelled and recorded. Later, at a meeting held on 18th November 1897, Maj.-Gen. Sir Robert Grant KCB, the Inspector-General of Fortifications, made the following suggestion.

> That the labels should have a broad arrow with a number underneath, and that the numbers should run right through consecutively, instead of each parish being numbered separately, as suggested by Mr. Squarey. This was agreed to.
>
> Mr. Squarey, in reply to the Financial Secretary, was unable to state exactly how many labels would be required, but probably between 800 and 900.[22]

Although it was originally proposed that these labels should be produced in the form of a small tile, the cost was considered to be too high and enamelled iron labels were adopted instead.[23] However, in some cases the numbers and the broad arrows were made of brass and some of these were still in existence in 1984, as for example on the thatched cottages at Hampshire Cross near the junction of Bulford Road and Lahore Road in Tidworth.

Those railway companies whose lines were in close proximity to the Plain, were now anxious to tap the sources of income which these new developments were providing by constructing new branches. The first of these was the Pewsey and Salisbury Light Railway which in 1897, the Great Western Railway Company was intent on building. The proposed line was to diverge from the railway running from Salisbury to Westbury at a point about three quarters mile west of Salisbury and to proceed through Durnford to the west of Amesbury. It was then planned to continue along the Avon Valley through Netheravon, Enford and Upavon before joining the Great Western's main line at Pewsey. The new railway which would run through the War Department's land for 4½ miles was strenuously opposed by the Committee since it was considered

> that the construction of this railway will neutralize, if not entirely destroy, the instructional value, in a military sense, of the River Avon. There is no military operation which can be more usefully practiced than the crossing of a river in the face of opposition. This practice the Avon will give; but if a railway is to be made alongside of the river, the defiles over or under that railway will not, as a rule, coincide with those over the river, and a situation will be produced which will practically prevent instruction being afforded.[24]

Rather more than two years later the Boer War had started and the British Army were to experience river crossings in the face of opposition under very different circumstances – over the Modder and Tugela rivers. Despite the opposition to the line the railway company decided to press its case, while the Committee felt that 'the War Department must oppose the construction of the line, lock, stock and barrel'.[25] This was done and on 20th April 1898 the Committee were informed that the Board of Trade had refused to approve the construction of 'a light railway upon land belonging to the War Department'.[26]

Another proposal was put forward by the Midland & South Western Junction Railway for a line from Ludgershall to Amesbury but the Committee were informed, at a meeting on 7th December 1897, that the War Department did not approve of the scheme and that it would be opposed.[27] However this opposition was later withdrawn 'as the proposed line ran entirely outside the boundaries of the War Department's property'.[28] Although this line as originally planned was never constructed, the branch from Ludgershall to Tidworth was completed in 1901. When however, a draft order for the proposed Amesbury and Military Camp Light Railway was considered by the Committee in December 1897 it was 'unanimously approved of'.[29] This line which ran from the London & South Western Railway's main line near Newton Tony to Amesbury was opened to the public in 1902 and extended to Bulford in 1906. Further information regarding these railways will be found in Chapter 13.

As the War Department came to be accepted as the new landlords, so neighbouring parishes and local interests began to turn to it for the financial help that had been provided by the previous owners. One of the first requests was received in November 1897 from the managers of the Enford schools since 'Sir E. Antrobus and Mr. Hussey Freke had now no interest in the schools on account of the purchase of their land by the War Department'.[30] This was followed by applications not only from other schools but also from several churches, district nurses, Salisbury Infirmary and the Netheravon Fire Brigade. Since the Committee had no powers to deal with these requests, the matter was referred to the Treasury which produced a comprehensive report laying down the policy to be followed.[31]

By the middle of 1898 the Salisbury Plain Committee had become deeply involved in the management of the new estate. This was marked by an apparent reluctance to delegate authority to the Agents to deal with those matters which occur from day to day on any large estate. It seems quite ludicrous that a committee composed of a parliamentary under secretary, two generals and one major-general had to authorize the expenditure of £2.8.od. (£2.40) for repairs to a keeper's dogcart.[32] Other examples of the tight control which the Committee exerted included such matters as deciding whether to buy an additional horse and cart for Tedworth gardens, dealing with boys who had been caught breaking windows and whether or not cottage chimneys should be swept.

By midsummer 1902, the War Department had acquired over 40,000 acres (16,185 hectares) and the Committee felt that an estate of this size warranted the services of a whole time resident agent. Consequently it was decided, at a meeting held on 2nd August 1901, to terminate the services of Messrs. Rawlence & Squarey who had been agents since 24th May 1897. This was to take effect on 31st December and a resident agent was to be appointed as from 1st January 1902.[33] At the next meeting it was decided to advertise the position on terms not in excess of the following scale of payments.

Salary	£450	per annum
Horse allowance	70	,, ,,
Lodging allowance (or house rent and tax free)	120	,, ,,
Office, office expenses and assistants	150	,, ,,
	£790	

In fact when the advertisement was drawn up, it stated that the salary was to be '£650 per annum to cover all expenses including horses, travelling, office expenses, assistants, etc.' with a house and office provided rent free or an allowance in lieu.[34]

On 19th January 1902 the Committee appointed Thomas Stirton as agent on these terms and he took up his duties on 11th February.[35] The names of his successors who have held the positions of War Department and Defence Land Agents from 1902 to 1985 will be found in Appendix II. Ablington House Figheldean which was demolished in 1963, was selected as the agent's house although part of it was used as the Estate

Office. In 1916 this office was moved from Ablington House to Brown's Farm, Durrington, where it remained until it was transferred to the present site adjoining Red House in 1920 or 1921, the exact year being in some doubt.[36]

It is usual for large estates to have a central depot known as an estate yard in which are situated sheds for storing building material and equipment, a carpenter's shop, paint shop, a covered area for machinery and so on. In some cases the yard includes a small sawmill and timber preservation plant. When the Estates Office was moved to the site next to Red House, an estate yard was established immediately to the south of it and this was still in operation in 1986 although the sawmill and timber preservation plant had been closed in 1960.

It was the original intention that the land which had been bought, should be used for training purposes, manoeuvres and camping and the Government was at pains to point this out. Speaking in the House of Commons on 22nd March 1897 the Under-Secretary of State for War said

> there was not the least intention of creating a second Aldershot on Salisbury Plain. The object was, in the first instance, to afford a manoeuvring ground for the troops, some of whom would probably belong to the Militia or the Volunteers. They should provide a range with the object of carrying out that portion of the training. There was no intention of erecting a great number of permanent barracks on Salisbury Plain, as had been done at Aldershot.[37]

Four months later, at a meeting of the Salisbury Plain Committee, he confirmed this statement although, perhaps, with a little less conviction.

> No buildings were at present contemplated at Salisbury Plain but it was impossible to say what the requirements might be in another 50 years.[38]

However, it would appear that by the end of 1898 the matter had been resolved since the Army Estimates for 1899–1900, contained the following statement.

> A scheme has been prepared for providing barracks which will be required by the additional troops now being raised ... The main feature of the scheme will be the erection of barracks for a considerable body of troops on Salisbury Plain, so that the fullest advantage may be derived from the large area of land recently acquired there for manoeuvring purposes.[39]

By the middle of 1899 the decision to erect buildings on the Plain had been made public and *The Devizes and Wiltshire Gazette* of 22nd June 1899 contained the following paragraph.

> Further information as to the barracks it is proposed to erect on the Government property on Salisbury Plain was forthcoming in the course of Mr. Wyndham's statement yesterday in moving a resolution in committee, preliminary to the introduction of the Military Works Bill. It is an important measure, being a continuation of the Military Works Loan Act of 1897 under which the Salisbury Plain property was purchased. Under the Bill

> the Government were asking with respect to barracks ... for £2,770,000 ... They were going to take £1,600,000 (of this) and spend it on Salisbury Plain. By doing that, he (Mr. Wyndham) believed they would give the taxpayer value for the money they had already sunk in that most desirable acquisition at Salisbury Plain ... As to concentration at Salisbury Plain, barracks would be built for seven battalions of infantry and six batteries of field artillery.[40]

G. Wyndham had been appointed Under-Secretary of State for War during 1898 in succession to W. St. J. Brodrick. A week later The Gazette published a further report 'from an expert's point of view' which included the remark that

> It is most satisfactory that the original intention not to build permanent barracks has been abandoned and that this locality is to become a large permanent military station.[41]

An account of the building and subsequent development of the military centres on the Plain will be found in those chapters which cover the areas concerned. In addition to erecting barracks there was also an urgent need to improve training facilities by the provision of rifle and artillery ranges. The steps which were taken to deal with these matters are described in the chapters on Bulford and Larkhill.

CHAPTER FOUR

Tidworth – The Early Days

During the past nine hundred years, Tidworth's name has undergone a number of changes. In the eleventh century it was Tedorde; two hundred years later it had become Thudeworth while in the fourteenth century it had changed to Tudeworthe. By the end of the next century it was spelt Tedeworth which is similar to the form used in Tedworth House today.[1] Although Speed used the form Tudworth,[2] Andrews and Dury on their maps of 1773 and 1812 which only marked the northern village, spelt it North Tidworth.[3] In 1817 the Ordnance Survey one-inch map showed the names of both the north and south villages, as well as the mansion, as Tidworth[4] and this spelling was retained on the six-inch map of 1886.[5] In spite of this, it would appear that in the nineteenth century the mansion and the estate and anything directly connected with them, adopted the name Tedworth and this occurs in Tedworth House and the Tedworth Hunt. The title page of the sale particulars of the estate issued by the auctioneers, Messrs. Daniel Smith, Son and Oakley in 1897, refers to 'the villages of South and North Tedworth' as does the conveyance to the War Department. However there would appear to be reasonable grounds for assuming that the usual form of spelling in the nineteenth century was Tidworth and this is supported by the following note which appeared in the *Andover Advertiser* in 1932.

> 'Tidworth' or 'Tedworth' has caused much controversy in its spelling since the barracks were built on Salisbury Plain. As readers know 'Tedworth' is the spelling retained by the Hunt but the villages and Garrison are now known as Tidworth and there is a common belief that the 'e' was always used until the soldiers came into the district. No so, however, for Mr. E.A. Sturgess of Faberstown has some old pewter pots in his possession bearing the names of former licencees and some are more than 100 years old. Each bears the modern spelling thus 'J. Moore, Ram Inn, Tidworth'.[6]

Tidworth is divided into two parts by the county boundary – North Tidworth which lies in Wiltshire and South Tidworth in Hampshire. The dividing line turns eastwards from Clarendon Hill, crosses to the south side of Lowa Road near the Garrison Theatre and then continues to the junction of Meerut Road and Lahore Road. At this point it bears to the north for about eighty metres before continuing eastwards to a point a little beyond the Globe Cinema and then on to the junction of Woodcock Close and Bourne Road. Here it turns to the north-east towards Pheasant Close and proceeds eastwards across Kennet Road to join the road to Perham Down. Although North Tidworth was the larger village, South Tidworth could claim to be more important in one respect since Tedworth House, the focal point of the neighbourhood, lay within the parish.

It is probable that considerably less interest would have been taken in Tedworth House had it not been for Thomas Assheton Smith who was not only a first-class horseman but has also been acknowledged as one of the greatest foxhunters of all time. Born in London on 2nd August 1776, he was brought up at Tedworth which belonged to his father and at the age of seven was sent to Eton where he was the youngest boy.[7] He continued at Eton for eleven years until he went up to Oxford in 1794. After completing four years at Christ Church, he settled in Leicestershire where his skill as a cross country rider soon became apparent and in 1806 he became Master of the Quorn. After ten seasons he moved to Lincolnshire where he took over the mastership of the Burton but in 1828, at the age of 52, he returned to Hampshire.[8] He established himself at Penton Lodge in Penton Mewsey, a little over a mile to the east of Weyhill where he formed a pack of hounds and started to hunt the surrounding country.[9]

In October 1827 Assheton Smith married and seven months later on the death of his father, he succeeded to the Tedworth Estate. Shortly afterwards he began the rebuilding of Tedworth House which took two years to complete (1828–30) and during this time he continued to live at Penton Lodge. In anticipation of moving to Tedworth he began to erect the stables and kennels in 1829. The stables which were built to his own design, provided boxes for 50 horses including hunters, carriage-horses and hacks.[10] In 1984 the building was in use as the District Works Office of the Property Services Agency, having been converted to offices some years earlier.

The original kennels were situated near the stables but after a short time they were moved to Lower Tidworth Farm in Humber Lane. In recent years this farm has been shown on the Ordnance Survey maps as Home Farm although the earlier name still appeared on an edition of 1961. The only part of the kennels that could be identified in 1984 was a building adjoining Humber Lane and although the internal walls had been removed it was still possible to discern the dished floors and outlines of the hound lodging rooms. Owing to the Second World War, hunting was discontinued and the hounds were distributed amongst several neighbouring packs which included the Wylye Valley and the Craven. At the end of the war another pack was formed by the new Master, J.C. Porter, who also provided kennels at Sunnyhill House, Milton, near Marlborough. The kennels were subsequently moved to Westcourt at Burbage, near Pewsey in 1957 and these were still in use in 1984.[11]

On moving to Tedworth, Assheton Smith turned his attention to one of the main problems when hunting the surrounding country – large areas of dense inaccessible woodlands, some of which were over 1,000 acres in extent. Although these provided unlimited cover for foxes, it was impossible to enter many of them in order to hunt hounds and he decided that the only way to deal with the problem was to cut straight rides through the woods. This involved the removal of any trees or underwood that were growing on the line of the proposed ride, grubbing out the stumps and levelling the ground. If Thomas Assheton Smith did not own a woodland which he wished to improve, he either rented it so as to lay out the rides himself or he used his powers of persuasion to encourage the owner to carry out the work. The clearance of these rides became so extensive and produced so much timber that Andover was said to have 'assumed the appearance ... of a great timber mart'.[12] All this effort and expense was repaid by some first-class runs and one of the best took place during the season of 1834–5 when a fox was found in an osier bed adjoining the Avon at Amesbury. On leaving the river, hounds hunted past Stonehenge to Barwick (Berwick St. James) and then on to Wishford where they crossed the Wylye and finally accounted for their fox in Grovely Wood, having covered 16 miles in 75 minutes.[13]

Although Assheton Smith is remembered as a great fox-hunter, he was also Member of Parliament for Andover for several years until 1832 when he opposed the Reform Bill and consequently lost his seat. In order to combine his sporting and political interests

> Mr. Smith used to hunt his hounds at Tedworth in the morning, and post in his light chariot with four horses, to Westminster in the evening.[14]

After he had voted in a division, he would return to Tedworth and hunt next day.

During 1830 a number of riots and disturbances took place in the neighbourhood and Assheton Smith was granted permission to raise 'a corps of Yeomanry Cavalry' which he did at his own expense. He was granted the rank of captain while the troopers were, for the most part, tenants on his estate or farmers living in the neighbourhood.

> They were viewed on one occasion in Tedworth Park by the late Duke of Wellington who spoke in high terms of their efficiency and soldier-like appearance. After the inspection and review, the troops were entertained at Tedworth Park.[15]

And so, some 65 years before the War Department commenced its acquisition of Salisbury Plain, there were troops based in Tidworth. A print showing Captain Assheton Smith carrying out an inspection at Andover in 1836 is reproduced on plate 5.

Mrs. Assheton Smith took a great personal interest in the tenants on the estate, the cottages 'being models of neatness and comfort' while 'the healthful and cheerful faces of the inmates well testified the care taken of them'.[16] In 1857 she built a school at the junction of the road to Perham Down (now Station Road) and the Collingbourne Ducis (now Pennings Road – A338).[17] Some idea of the extent to which this locality has changed, can be obtained from the illustration on page 00 which shows the school before the surrounding land was developed for building. The photograph was taken at the junction of what is now Lahore Road and Meerut Road looking down Lahore Road, the woodland on the skyline being Furze Hill. The school continued until 1980 when it was closed and although it was still standing in 1984 it was then in a very dilapidated condition.

Thomas Assheton Smith died on 9th September 1858 at the age of 82, his wife dying eight months later on 18th May 1859. He was an extremely wealthy man who was quick-tempered but kind-hearted and since his life revolved around hunting to the exclusion of much else, he became something of an eccentric. If he did not finish a day's hunting near his trap or other vehicle, he rode home at full gallop all the way. George Carter, his huntsman, recalled this in later years.

> He would come along, waving his hand, and I'd say to Cowley, the whip, 'We must get out of the way or Mr. Smith will be right over us' and he'd come past as hard as he could gallop, never taking any notice of us or the hounds. In frosty weather he used to have three or four horses out, and gallop them up and down as hard as ever they could lay legs to the ground and I don't believe he would have pulled up whoever got in his way ..... Mr. Smith he were an odd man; a very odd man.[18]

Since there were no children of the marriage Mrs. Smith left the Tedworth Estate to her favourite nephew and godson, Francis Sloane Stanley, who inherited the property on her death in 1859.[19] As Stanley was living in Leicestershire, he decided to find a tenant for Tedworth House and in 1860 it was let to Edward Studd who had recently returned from India. Studd was a keen hunting man who soon became interested in racing and decided to set up a training establishment. No doubt he was attracted to Tedworth by the exceptionally fine stables and it was not long before he had twenty horses in training and had built a small racecourse in the Park. His enthusiasm and determination were rewarded in 1866 when he won the Grand National with Salamandar, while in 1871 another of his horses – Despatch – came second. When, later on, Salamandar fell at a fence and broke a leg, so that he

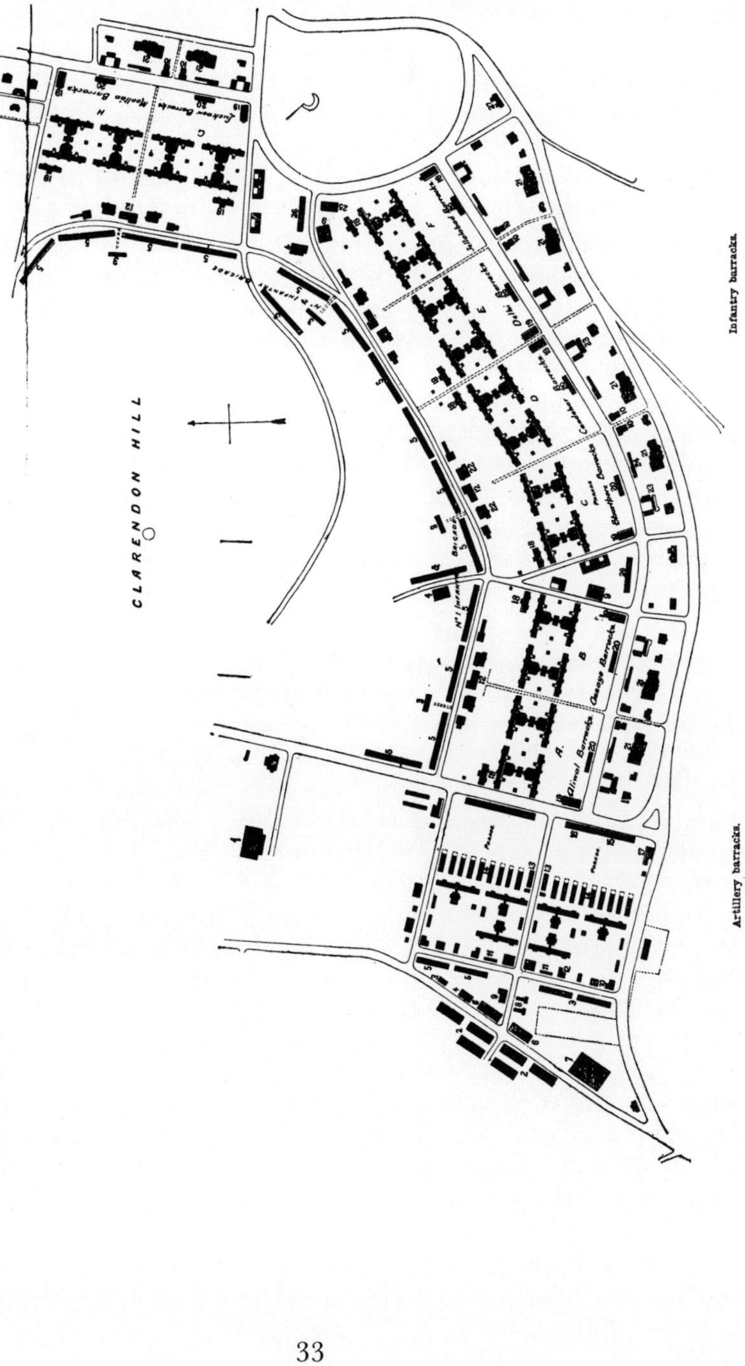

3.THE NEW BARRACKS AT TIDWORTH
This plan was published in *The Devizes and Wiltshire Gazette* 27 December 1900

had to be destroyed, the whole family came down to dinner that evening, dressed in black.[20]

When Studd was on a visit to London in 1876, a friend persuaded him to go to a meeting to hear one of the American evangelists, D.L. Moody, who, with I.D. Sankey, was touring this country. This changed his whole life and he resolved to follow in Moody's footsteps. He gave up racing, disposed of his horses and replaced the furniture in the hall at Tedworth House with chairs and benches so that prayer meetings could be held on Sunday evenings and to these he invited his neighbours.[21]

In 1878 when he was on his way to attend one of Moody's meetings, he remembered that he had intended to bring one of his grooms. Stopping the carriage, he ran back to the house but the exertion was too much for him, so that he burst a blood vessel and died very shortly afterwards.[22] His third son, C.T. Studd, was an outstanding cricketer who played for England when the Australians beat the English eleven for the first time. He was subsequently in the team which went to Australia and brought back the Ashes.[23] Like his father he was greatly influenced by Sankey and Moody and in 1885 went as a missionary to China.

On 11th December 1876, Stanley sold the Tedworth Estate to Sir John Kelk, Bt.[24] and as this was approximately two years before Edward Studd's death, the sale was presumably subject to his tenancy as Kelk did not take possession until after Studd had died. John Kelk was a well-known civil engineering contractor who specialized in the erection of large buildings. He was closely connected with the constructional work for the Great Exhibition of 1851; he undertook the erection of the Albert memorial in 1864 'without pecuniary benefit'; he was jointly responsible for the construction of Millwall Docks in 1868 and built Victoria Station in 1858–1860 and Smithfield Market in 1866–1869. He was Member of Parliament for Harwich from 1865 to 1868 and was created a baronet in 1874.[25] He built St. Mary's Church which stands beside the Tedworth–Shipton Bellinger road (A338) the foundation stone of which Lady Kelk laid on 29th April 1879. On his death, he was succeeded by his son John William Kelk who sold the Tedworth Estate to the War Department in 1897. Two reminders of his ownership are still be be seen in Tidworth. The first is a row of three flint and brick cottages on the Ludgershall road, about 80 yards to the east of the Ram Inn, the centre one bearing a stone inscribed 'JWK 1890'. The second is the reading room and attached cottage which he built close to the junction of what is now known as Lahore Road and Bulford Road. He died on 22nd March 1923. In 1984 there were still a number of buildings in Tidworth which had been built before the Estate was sold to the War Department.

After the purchase of the estate had been completed, little time was lost in pressing ahead with the planning and design of the new barracks.

The site which had been selected, lay at the end of a shallow valley or re-entrant which followed the eastern and southern slopes of Clarendon Hill, just below the 400ft. contour. Bazaar Road which follows this contour, rises from 393ft. above sea level at its junction with Agra Road to 406ft. at Margha Road, falling to 395ft. where it meets Nainital Road. On the southern side of the valley, the 400ft. contour runs from a point a little to the west of Aliwal Officers' Mess, eastwards across Bulford Road about 100 metres north of the junction with Humber Lane. From there, it bears to the south-east as far as the south-western corner of Tedworth Stables and then turns south-westwards through the garden of Clive House. The lowest point in this valley is at the junction of Dasna Road and Grand Trunk Road (359ft.) while the highest is the hill on which St. Michael's Church stands (404.5ft.).

In December 1900 *The Devizes and Wiltshire Gazette* published a plan of the proposed development which had been made available by the Commander Royal Engineers, Col. J.M. Barklie who was responsible for the supervision of the work.[26] A comparison of this plan which is reproduced on page 00, with the 25-inch Ordnance Survey of 1936, shows that the layout of the roads and buildings was basically the same as that which was finally adopted, with two major exceptions. Firstly, in addition to the eight infantry barracks which were to be built, two artillery barracks were to be constructed immediately to the west of Aliwal. These were both the same size and similar in layout and occupied a site between what would have been westward extensions of The Mall and Bazaar Road. Secondly, St. Michael's Avenue was omitted and instead Kirkee Road followed a right handed curve to join Lowa Road at a point near its present junction with St. Patrick's Avenue.

Each of the eight barracks were designed to accommodate one infantry battalion: Aliwal, Assaye, Bhurtpore and Candahar being allocated to No.1 Infantry Brigade and Delhi, Jellalabad, Lucknow and Mooltan to No.2 Infantry Brigade. The two artillery barracks were each intended to hold one brigade, of three batteries, of field artillery but no provision was made for any cavalry barracks and this was the subject of some comment.

> When we wrote in June last year there was some question as to whether Cavalry Barracks would or would not be in the Scheme. Speaking at the dinner of the Wiltshire Society a week or two earlier the Marquess of Lansdowne, then Secretary of State for War, said provision would be made for a Brigade of Cavalry. But speaking in the House of Commons a little later Mr Wyndham, Under Secretary of State, referred only to Infantry and Artillery and had nothing to say as to Cavalry. As the plans now at Tidworth House show, no Cavalry Barrracks are in the Scheme.[27]

The plan of the proposed development did not extend northward beyond Plassey Road nor did it show all the buildings which were to

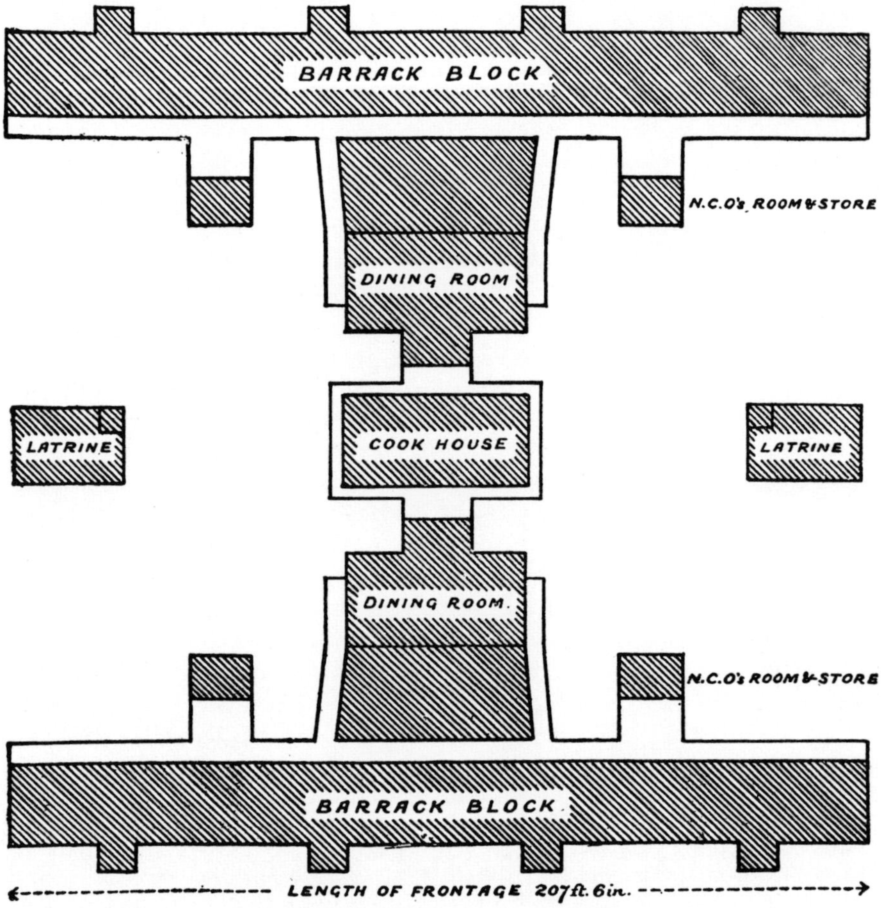

4. THE NEW BARRACKS AT TIDWORTH
Ground plan of one of the eight
infantry blocks published in
The Devizes and Wiltshire Gazette
27 December 1900

be erected. 'There are besides, others to accommodate Royal Engineers, Army Service Corps and Army Ordnance Corps'.[28] An Ordnance Survey Map showing land belonging to the War Department in 1903 marks the approximate locations as follows.[29]

'R.E.Bks' – to the north of Zouch Farm between what is now Zouch Close and Sidbury Circular Road.
'A.S.C.Bks' – on either side of Station Road.
'A.O.D.Bks' – along the line of Ordnance Road.

It was also reported that the whole scheme would cover some 460 acres (186 hectares), the infantry barracks alone occupying 220 acres (89 hectares).

In the infantry barracks accommodation was provided in two identical blocks each being divided into two sections consisting of sleeping quarters – the 'barrack block' – and a dining room. One central cookhouse served the two dining rooms as shown in the drawing on page 00. The sleeping quarters occupied two floors with eight rooms on each, communication between the floors being by means of an external staircase and an iron balcony running the full length of the building. The cookhouse and dining rooms were only single storey buildings.[30] The limits of the infantry barracks were formed by Bazaar Road on the north and west and by Grand Trunk Road and St. George's Road on the south and east. Above Bazaar Road were the soldier's married quarters while on the opposite side of Grand Trunk and St. George's roads were the officers' messes and, except for Lucknow and Mooltan, the Commanding Officer's residences.

It is not known when it was decided to alter the original proposals but by the time work on the barracks had begun, several fundamental changes had been made. The construction of the artillery barracks had been abandoned while the allocation of the eight barracks was amended so as to divide them equally between cavalry and infantry as follows

Cavalry: Aliwal, Assaye, Candahar, Mooltan.
Infantry: Bhurtpore, Delhi, Jellalabad, Lucknow.

The married quarters in Bazaar Road were built in sets each consisting of two rows of 24 houses, sited one behind the other instead of being in single rows extending the whole length of Bazaar Road, as was originally intended. Two rows were allocated to each barracks and named accordingly, as for example 'Bhurtpore 1 to 24' and 'Bhurtpore 25 to 48'. By 1984, a large number of modern married quarters had been built elsewhere in Tidworth and the Assaye and Jellalabad sets and half of the Candahar set (25–48), had been demolished while most of the remainder were unoccupied and in a derelict state. They were known locally as 'Merthyr Tydfils' presumably since it was thought that they bore some resemblance to the rows of houses to be seen in some Welsh mining areas. Every set of quarters was provided with two laundries –

one for each row – which were sited behind the second row, but by 1984 these had either been demolished or put to some other use. The pair behind Jellalabad had become part of the Isolation Hospital and now forms Nightingale Lodge while those behind Candahar had been converted into unit offices.

Another change in the original plans was the construction of St. Michael's Avenue between Grand Trunk Road and Lowa Road, in order to provide a direct link between the main barrack block and Lucknow and Mooltan Barracks. It also facilitated the building of the section of the camp railway between Kirkee and Lowa Roads. This line which ran from Tidworth Station through each of the barracks was completed during 1901 when the railway from Ludgershall to Tidworth came into use for moving men and materials which were required for the building of the barracks. In order to obtain the necessary work force required for the construction of the barracks, numerous other buildings and the camp railway, it was necessary to recruit labour over a large area. So as to overcome the problem of housing a corrugated iron hutted camp was built at Brimstone Bottom midway between Tidworth and Ludgershall. Officially known as the Navvy Village and shown by this name on the Ordnance Survey map of 1903, it soon became known as 'Tin Town' on account of its construction. Although a certain number of facilities were provided including a church, there were not many opportunities for relaxation and the *Andover Advertiser* suggested that

> A regular train from Andover about 10.15 each night would be appreciated by many on whom the beauties of Tin Town and Ludgershall have somewhat palled.[31]

In April 1903 a temporary siding was laid so that workmen could travel to Tidworth by train[32] and at the same time the railway was used to carry the large quantities of building materials which were needed.

By the end of 1900 the sloping ground around the base of Clarendon Hill was 'plotted out with marks of the surveyors' and it was reported that 'the necessary preliminaries to the placing of the contracts are actually completed'.[33] In Tidworth, building began in 1902[34] and *The Barrack Book 1929*[35] gives the 'year of construction' for many of the original buildings in Tidworth Garrison. This shows that the first buildings to be erected, although not actually in Tidworth, were those which comprised the Isolation Hospital at Brimstone Bottom which was built in 1900. It was sited between the Tidworth–Ludgershall road and the railway line, immediately to the south of Tidworth Down Farm. In 1903 the sewage works in Tedworth Park were constructed and the Tidworth Military Cemetery was laid out to the north of The Pennings. It would seem very probable that the first buildings to be completed in the barracks area were ten houses in Grand Trunk Road known as Clarendon Terrace which are situated between Agra and

Baroda Roads. These were built in 1904[36] so as to provide accommodation on the construction site for a limited number of War Department employees and were subsequently allocated as Warrant Officers' quarters.[37]

The construction of an entirely new military centre was a considerable undertaking and was bound to occupy some time but it was felt in some quarters that the work was not proceeding very quickly. In September 1903 *The Andover Advertiser* observed that 'at Tedworth the progress of the new permanent barracks has been painfully slow to the uninitiated'.[38] However, *The Barrack Book* shows that the 'year of construction' for all the barracks was 1904 except in the case of Assaye which is given as 1905. In view of this, it is interesting to note that the dates which appear over the officers' messes are 1905 except for Aliwal which is 1904. It could be assumed from this that, except in the case of Aliwal and Assaye, the messes were not finished until a year after the barracks had been built, that is, if the 'year of construction' indicates the year in which a building was completed. Another set of buildings which were erected in 1904 were those of the Station Veterinary Hospital and the priority given to its construction was very understandable in the days when the horse played such a vital part in the British Army. The Hospital was situated to the north of Aliwal Barracks about 120 yards from the end of Bazaar Road and was finally closed and demolished in 1950.

The layout of the roads in the barrack area is based on three roughly parallel lines which ran from Aliwal in the west to Mooltan in the north. These are, Bazaar Road which follows the shoulder of Clarendon Hill and to which reference has already been made earlier in this chapter; Grand Trunk Road with its continuation by way of St. Michael's Avenue and St. George's Road running below Bazaar Road, and The Mall, Bulford Road and Meerut Road forming the southern and eastern boundaries. Originally the military railway ran between and parallel to, Bazaar Road, Grand Trunk Road and St. George's Road but when the railway was lifted after the Second World War, the track bed was converted into another road. These roads are joined by a series of connecting roads which have been named after towns and villages in India, in alphabetical order beginning with Adampur Road in the west to Nainital Road to the north of Mooltan Barracks. Subsequently Nepaul Road and Plassey Road were added. (A detailed list of these roads will be found in Appendix III together with the locations of the places after which they are named.)

The eight new barracks were also named in alphabetical order, in commemoration of battles or sieges of which six had taken place in India and two in Afghanistan and these are described in the following summaries.

Aliwal

The village of Aliwal is situated in the Punjab close to the southern bank of the river Sutlej being about twenty-five miles due south of Jullundur and sixteen miles west of Ludhiana. During the First Sikh War of 1845–1846, a British force under the command of Major-General Sir Harry Smith, a veteran of Waterloo and the Peninsular War, was in contact with a Sikh army lead by Ranjur Singh. In the initial stages, Smith intended to mount an attack on the fortress of Baddowal but before he was able to do so the Sikhs withdrew towards the Sutlej.

With some twelve thousand men and thirty-two guns under his command he eventually moved forward on 28th January 1846 and after covering some eight miles or so, reached the village of Porrain. This stood on a ridge overlooking a small grassy plain about a mile in width, beyond which lay two villages, Aliwal which was fortified, and Bhundri, while between them ran a line of earth-works.[39] As the British force entered Porrain, the Sikh columns numbering some fifteen thousand men, came into view on the plain below but realizing that the British had found them, they turned and occupied the two villages. Deploying his troops

> Smith advanced as if on parade ... The sun shone brightly; there was no dust; every manoeuvre was perfectly executed; and the twelve thousand men acted as one. Rarely has there been a more stately prelude to a general action.

The first shots were fired by the Sikh artillery which enabled their position to be observed and Smith decided first to take Aliwal and then to attack the centre and left. He also appreciated the fact that the Sikh commander had made a grave tactical error in selecting a position behind which a river flowed since this could seriously prejudice any attempt to retreat.[40] Although the Sikhs fought with great gallantry, the admirably co-ordinated use of cavalry, artillery and infantry by Sir Harry Smith proved too much for them while in their efforts to withdraw, the river claimed many lives. The casualties suffered by the British force amounted to five hundred and eighty-nine of whom one-third were killed, while the Sikhs admitted to a loss of three thousand dead.[41]

Assaye

This battle which occurred during the Second Mahratta War (1803–1805), took place near the village of Assaye which lies some 45 miles north-east of Aurangabad and 75 miles south-west of Akola in present day Maharastra. It was fought on 23rd September 1803 between the British forces commanded by Sir Arthur Wellesley (afterwards The

Duke of Wellington) and the Mahrattas under Prince Sindia. The strength of the opposing force has been estimated at between forty and fifty thousand which was comprised of twenty to thirty thousand cavalry, ten thousand infantry, a large number of irregulars and at least a hundred guns. The British and native troops numbered about seven thousand infantry and cavalry, fourteen guns and eight light 'galloper-guns' which were attached to the cavalry.[42] The battle took place between the rivers Kaitna and Juah, Assaye standing between these two rivers on the southern bank of the Juah. Although Wellesley gained an initial success by finding a ford across the Kaitna, contrary to the information given to him by his guides, the fighting was fierce and relentless. The Mahratta army fought with great courage and determination but by the evening, it no longer existed as a fighting force. The British losses amounted to 'nearly six hundred and fifty Europeans and eight hundred natives killed, wounded and missing ... The enemy left twelve hundred dead on the field; their wounded were estimated at four times as many; and they abandoned to the British ninety-eight guns'.[43]

It is possible that the reason why Assaye was chosen as the name of the first of the new barracks can be found in Fortescue's comment that '...though every man, British or native, played his part with superlative gallantry, Assaye presents a roll of valiant deeds which is unsurpassed in our military history'.[44]

Bhurtpore

The Second Mahratta War was, to a large extent, fought in two separate localities some 500 miles apart. The southern area which lay between Aurangabad, Burhanpur, Akola and Parghani included Assaye while the northern area extended eastwards from Jaipur to Cawnpore and northwards to Delhi and Jullundur. In this northern locality lay Bhurtpore (now Bharatpur) about 100 miles east of Jaipur and 35 miles west of Agra. While the campaign in the south was conducted by Sir Arthur Wellesley, the operations in the north were under the command of General Gerard Lake (afterwards Lord Lake).

On 2nd January 1805, Lake reached the outskirts of Bhurtpore, a town of considerable size having walls over eight miles in circumference and within which stood a massive fort.

> The whole of this vast perimeter was surrounded by a mud wall of great height and enormous thickness, with round bastions, mounting innumerable guns, at short intervals, and by a very wide and deep ditch ... The whole lay within a belt of jungle and swamp, the water of which could be drained off to fill the ditch. Indeed upon the arrival of Lake's army a large expanse of water on the north-west side suddenly disappeared, and not for some time was it discovered that the whole of it had passed into the ditch.[45]

The first attempt to take the town was made on 9th January but this proved unsuccessful and resulted in heavy losses. Further attacks were launched on 21st January and 20th and 21st February but each attempt to breach the walls failed, the net result of these abortive operations being a total of 3,000 casualties in the British force.[46] Failure to achieve success was mainly due to insufficient preparations being made for the various attacks, to Lake's impatience and to a lack of men, stores and equipment. To make matters worse, the mud walls which it was thought, could be breached by gunfire, proved to be virtually insurmountable; shells buried themselves in the soft material and even when a portion collapsed, it was reduced to a thick powder over which a man could not walk without sinking into it.

On 24th February, Lake withdrew in order to prepare for another assault but the strain of being under continuous attack had begun to tell and on 10th March 1805 the rajah of Bhurtpore sued for peace and a treaty was concluded.[47] This was conscientiously observed by both parties but by 1823 the rajah, Baldeo Singh, was old and in poor health and he wished to ensure that his son Bulwant Singh who was only six years old succeeded him. There were however, indications that his nephew, Durjan Sal, might attempt to seize power from the boy and the old rajah accordingly placed his son under British protection. A year later Baldeo Singh died and the nephew entered the town by force and took the son prisoner.[48] Such a state of affairs could not be tolerated and appropriate action was taken. The first step was to occupy the swamps and get control of the sluices which admitted water for the protecting ditches and this was achieved on 10th December 1824. On 23rd December the seige batteries opened fire and during the next three weeks mines were driven to the foot of the walls. On the morning of 18th January 1825, the mines were detonated and the assault began and by the late afternoon the town had surrendered.[49] Durgan Sal and his family were captured and the legend of Bhurtpore as an impregnable fortress had run its course. The British losses amounted to about 1,000 while those of the opposing forces numbered some 14,000.[50]

Candahar

This Afghan city lies some two hundred and eighty-five miles south-west of Kabul and one hundred and thirty miles north-west of Quetta. The name 'Candahâr' which is more frequently spelt 'Kandahar', became prominent during the First Afghan War in 1839–1842 and the Second Afghan War of 1878–1880 when the city was relieved by Maj.-Gen. Sir Frederick Roberts VC (afterwards FM Lord Roberts). In the battle honours of those regiments that took part in the first operation, the name appears as 'Candahar 1842' while for those that were involved on the second occasion, the honour appears as 'Kandahar 1880'. It could be deduced from this that the name of the Tidworth barracks which is spelt with a 'C' refers to the campaign of 1842.

In January 1842 when Maj.-Gen. Sir William Nott was in command of the troops at Candahar, a general uprising took place in the province. Subsequently on 7th March, Nott moved out of the city in an attempt to force a confrontation. In this he was unsuccessful as immediately his troops came within range of the tribesmen, they withdrew and so successfully enticed him further from his base. After four days he returned to Candahar, only to find that the city had been attacked in strength during his absence but that this had been successfully repulsed.[51] Later in May, Nott made another sortie from Candahar and defeated a force of about eight thousand Afghans that had taken up a position a mile from the walls of the city.[52] In the meantime it had been decided to withdraw the British forces from Afghanistan and Nott who was to move to Kabul, left Candahar at the end of August and arrived at his destination in mid-September. On 12th October the army began its march from Kabul to Peshawar which it reached in November so concluding the First Afghan War.

On 27th July 1880, a British force of about two thousand was defeated at Maiwand in southern Afghanistan and shortly afterwards the garrison in Kandahar, commanded by Maj.-Gen. James Primrose, came under siege. To deal with the situation troops under Maj.-Gen. Sir Frederick Roberts, VC were despatched from Kabul and after a march of some three hundred miles they reached Kandahar early on 31st August. On arrival the Afghans were found to be occupying a steep ridge about two miles north-west of the city in considerable strength. Roberts decided to attack on 2nd September and he did so with such vigour that the enemy were unable to hold their position and retreated in complete disorder.[53] This success marked the close of the Second Afghan War.

Delhi

The siege of Delhi which took place during the Indian Mutiny of 1857–1859 was a matter of considerable concern since the city occupied a position which was both strategically and politically important. The cause of the Mutiny has frequently been attributed to the introduction of a new type of cartridge which required the end to be removed before loading. This was done by biting it off and as it was rumoured that the cartridge had been treated with cow and pig grease, great offence was given to both Hindus and Moslems on religious grounds since these animals were considered to be 'unclean'. There were in fact a number of other reasons including poor relationships between the officers and their men as well as Government action which had undermined the morale and discipline of the Indian troops.

On 10th May 1857 a revolt occurred in two native infantry regiments at Meerut and on the following day the mutineers marched on Delhi where they attacked the palace and murdered all the British subjects whom they could find. To deal with the situation, a British Force under General George Anson left Ambala on 25th May but the next day when some 85 miles from Delhi, Anson was struck down by cholera and died within a few hours. Under his successor, Gen. Sir Henry Barnard, the force reached Delhi ridge a mile from the city where it took up its position,[54] and on 19th and 20th June the enemy attacked in great strength but were driven back after heavy fighting. For the second time in six weeks cholera claimed a Commander-in-Chief when Sir Henry Barnard died on 5 July after only a few hours illness. His place was taken by Gen. Reed but twelve days later his health failed and he handed over command to Brig.-Gen. Archdale Wilson.[55] After receiving reinforcements, the attack began on 14th September and a footing was obtained in the city. Heavy fighting continued but after six days there was little resistance and the King's Palace together with the Mosque of Jama were occupied. On 21st September 1857, Delhi was once again under British rule and its capture has been described as 'a feat of arms most creditable to the courage and endurance of the troops'.[56]

Jellalabad

In 1841, the town of Jellalabad which is situated about seventy-five miles east of Kabul, lay for the most part on the south side of the Kabul river, its defences being in a very dilapidated condition. On 12th November of that year, Brig.-Gen. Sir Robert Sale entered the town and proceeded to repair the walls and improve the defences. On the following morning the Afghans launched the first of their many attacks. Sale considered that he ought to remain at Jellalabad until a relieving force could be sent to help him and the arrival of Dr. Bryden, the sole survivor of the disastrous retreat from Kabul, on 12th January 1842 did not encourage him to change his mind. Eventually, after numerous setbacks, a force under Maj.-Gen. George Pollock, reached Jellalabad on 16th April 1842 and relieved the garrison.

Lucknow

In September 1857 a British force was assembled at Cawnpore under Gen. Henry Havelock with the object of relieving the besieged garrison at Lucknow. On 16th September Gen. Sir James Outram joined Havelock and although his senior, he generously announced that Havelock would continue to be in command. The march to Lucknow – some fifty miles to the north-east – began on 18th September and five days later Alam Bagh, a large walled garden two miles south of Lucknow, had been captured.[57] On 25th September, the British force entered Lucknow

(T. L. Fuller & Sons)

1. Washing sheep at Enford Manor in about 1902. In the background is the Manor House with a cottage adjoining it on the left. The house has since been demolished and only the cottage remains.

HANTS AND WILTS.

PARTICULARS WITH PLAN

OF THE

TEDWORTH PARK ESTATE

A HIGHLY IMPORTANT

FREEHOLD RESIDENTIAL PROPERTY,

COMPRISING AN

Exceptionally Fine Mansion,

FINE WELL TIMBERED PARK & PLEASURE GROUNDS,

NUMEROUS COMPACT FARMS,

WITH EXCELLENT FARM HOUSES AND MODERN HOMESTEADS;

OVER 750 ACRES OF WOODLANDS & PLANTATIONS,

Forming fine Game Coverts, for which the Estate is noted; together with the

KENNELS OF THE WELL KNOWN TEDWORTH HOUNDS

The whole of the Villages of South and North Tedworth, and part of Shipton Bellinger,

FORMING ALTOGETHER A FINE

SPORTING AND RESIDENTIAL DOMAIN,

Extending over an area of

6,618 ACRES,

Situated in the Parishes of NORTH and SOUTH TEDWORTH and SHIPTON BELLINGER, 3 miles from Ludgershall Station on the Andover and Swindon Branch, and 5 miles from Grateley on the Main Line of the South Western Railway, whence London, which is 74 miles distant, is reached in 2 hours; about 15 miles north of Salisbury, and about the same distance south of Marlborough.

FOR SALE BY PRIVATE CONTRACT

BY

Messrs. DANIEL SMITH, SON & OAKLEY.

Further Particulars and Plans may be obtained of Messrs. DANIEL SMITH, SON & OAKLEY, Land Agents, Surveyors and Auctioneers, 10 Waterloo Place, Pall Mall, London, S.W., where Views of the Mansion and Park may

2. The cover of the Sale Particulars of the Tedworth Park Estate which was bought by the War Office on 29th September 1897.

(*Daniel Smith*)

(*The Army & Navy Illustrated*)

3. Transport passing the Poultry Cross in Salisbury during the Army Manoeuvres of 1898.

THA

MILITARY MANOOVERS

IN THA

Nayberhood a Zalsbury

ZEPTEMBER, 1907,

BY

Measter Benjamin Sloper.

Being an account of the

VARIOUS OPERATIONS.

also the

RECEPTION of the WILTSHIRE REGIMENT

by the

CITY OF SALISBURY.

PRICE SIXPENCE.

SALISBURY.

R. R. EDWARDS, 6, CASTLE STREET,

AND ALL BOOKSELLERS.

4. The cover of the booklet by 'Benjamin Sloper' (Edward Slow of Wilton) which described the manoeuvres of 1907.

(*C. J. J. Berry*)

5. A view of Andover High Street with Captain T. Assheton Smith inspecting the 'corps of Yeomanry Cavalry' which he had raised. The date is about 1836.

(*L. E. Trevett*)

6. A photograph of Hampshire Cross, Tidworth taken in 1913 or shortly afterwards, from the western end of what is now Lahore Road. The thatched cottages and the pair on the right were still in use in 1986.

(*L. E. Trevett*)

7. The Post Office, North Tidworth about 1913. The road beyond the walking figure is now Pennings Road and on the right is the road to Ludgershall.

(*N. D. G. James*)

8. One of the original barrack blocks in Assaye Barracks which were built in 1905, showing the outside balcony which gave access to the first floor. Photographed in 1984.

(*N. D. G. James*)

9. The original married soldiers' quarters known locally as 'the Merthyr Tydfils' of which each barracks had two blocks of 24. This photograph shows 'Lucknow 25–48' situated on the west side of Bazaar Road.

(*Mrs. H. E. P. Colley*)

10. Tidworth Church of England School which was built by Mrs. Assheton Smith and although very dilapidated was still standing in 1985. The photograph was taken at Hampshire Cross, looking down what is now Lahore Road towards the present day Station Road. The date is probably about 1895.

(*N. D. G. James*)

11. The original Army School which was built in 1909 in Bazaar Road opposite the Garrison Theatre.

(*N. D. G. James*)

12. St. Andrew's Presbyterian Church at the junction with Kirkee Road and St. Patrick's Avenue. It was the first church to be built in Tidworth which was opened in January 1909 and closed in 1964.

(*N. D. G. James*)

13. The Church of England Garrison Church of St. Michael which was dedicated on 4th May 1912.

and, after heavy fighting, reached the Residency which they succeeded in entering. However, although they were inside the city, they were not in command of it, and fighting continued without respite. After finding a large store of grain in the Residency, Outram decided to stay in Lucknow although the medical stores and food had been left at Alam Bagh. His action was, in effect, to reinforce the garrison but not to relieve it.[58]

At the end of October another column was assembled at Cawnpore which included a Naval Brigade and on 9th November, Sir Colin Campbell who had recently been appointed Supreme Commander in India, joined the force. Three days later he reached Alam Bagh and proceeded to occupy Dilkusha which lay on the outskirts of the city and which he fortified as a depot.[59] On 16th November the advance continued and by the next evening the British force controlled the area between the city and the Gumti river and had gained access to the Residency. The rest of Lucknow was still in rebel hands and after making good the line of retreat, canvas screens were erected across any open ground that was in view of the enemy. Two days later under cover of these, the women and children with the sick and wounded were moved from the Residency to Dilkusha while harassing fire was directed against all enemy strong points. Finally, at midnight on 22nd November, the garrison withdrew in complete silence and by daybreak all were safely at Dilkusha.[60] Two days later Havelock died from dysentry.

Having successfully completed the relief of Lucknow, Campbell's attention was directed to Cawnpore where the garrison was under constant attack. Consequently it was not until February 1858 that preparations were made for recapturing Lucknow. Eventually, with a force of some 30,000 men he began the advance on Lucknow on 2nd March and on reaching Dilkusha, built two bridges across the Gumti.[61] Outram who had been holding Alam Bagh moved forward, crossed the river and reached the Iron Bridge opposite the Residency. By 14th March the Kaiser Bagh Palace had been taken and a week later all resistance had been overcome. However, many of the mutineers escaped largely due to Campbell's refusal to allow Outram to cross the Iron Bridge and cut off their retreat.[62]

Mooltan

Mooltan which is more commonly spelt Multan, is situated in the Western Punjab of Pakistan, some 190 miles south-west of Lahore. Although the Treaty of Lahore which had been signed on 8th March 1846 marked the end of the First Sikh War, the situation had remained unsettled. It finally came to a head on 20th April 1848 when two British officials, Mr. Vans Agnew and Lt. Anderson who were resident magistrates at Mooltan, were murdered and it was this event that lead to the beginning of the Second Sikh War of 1848–1849. Action was immediately taken by Lt. Herbert Edwardes, a young officer in the Bengal Native Infantry, who after defeating the Sikh leader, Dewan Mulraj forced him to retreat to Mooltan. At this point Gen. William Whish who was commanding a British force, was ordered to Mooltan which he reached on 18th August 1848 being joined by additional troops on 4th September.[63]

An attack which was launched on the Sikh position on 10th September failed, but two days later Whish mounted another assault which drove the enemy back. Shortly afterwards news was received that another Sikh leader, Shere Singh, had joined Mulraj and as the enemy's strength was now infinitely greater, Whish withdrew.[64] In October Shere Singh left Mooltan to join other Sikhs further north and after some delay Gen. Whish returned to the town and carried out a preliminary attack on 7th December. The main operation which began on 27th December continued until 22nd January 1849 when Dewan Mulraj surrendered.[65]

CHAPTER FIVE

Tidworth – Development and Expansion

In the original plans which had been drawn up for the new Barracks there was one notable omission – no provision had been made for a hospital. Although in 1900 an isolation hospital had been built at Brimstone Bottom[1] this did not deal with the non-infectious ailments and casualties which inevitably occurred in a large garrison. The absence of a hospital at Tidworth may have been due to the fact that one of 200 beds had been established at Bulford in 1905[2] but as more troops took over the new barracks the provision of a Tidworth hospital became a matter of considerable urgency. It was clear that the quickest results would be obtained by converting one of the barracks into a hospital and it is probable that Delhi was chosen on account of its central position. In July 1907, Lieut.-Col. C.E. Faunce who had commanded No. 20 Company RAMC at Bulford from 1904 to 1905,[3] was appointed Officer-in-Charge of the 'Military Hospital Tidworth'[4] and it may be assumed that the Hospital came into being on that date. However it would seem that the adaptation of Delhi was only intended to be a temporary arrangement and that the erection of a new hospital was under consideration. When the Matron-in-Chief of Queen Alexandra's Imperial Military Nursing Service inspected the hospital on 20th May 1908 she reported that

> No pains have been spared to render the barrack rooms like hospital wards, and to make the patients as comfortable as possible but it is more than evident that nothing short of a new hospital is required. The number of beds is 120, but these are insufficient for such a large garrison.[5]

There were at that time nineteen wards: seven surgical, nine medical and three officers' which were run by two sisters and three staff nurses who were about to be supplemented by another nurse.

Referring to the buildings, the Report stated that the ventilation was 'very good' – far too good it would seem, since

In the winter it is most difficult to keep the temperature of the wards even as high as 40° Fahr. The ceilings are low and have to be constantly whitewashed as the hanging lamps blacken them. This hospital is only composed of barrack rooms and the resulting accommodation most inadequate ... No hot water laid on, all has to be carried from the kitchen.[6]

Some items of furniture and equipment were not up to standard and had probably been discarded by other hospitals.

The equipment is well looked after but some of it is very old such as the bedside lockers. Covers for the arm-chairs have shrunk and become too small to use. The new coal scuttles ... are poor and shabby-looking articles ... The dressing-wagon sent from Woolwich for the surgical ward is an old-fashioned thing, roughly repainted and not at all suitable.[7]

The Matron-in-Chief also visited the Military Families Hospital on the same day but it is not clear whether this was part of the main hospital or not. It was very small, having only six beds and her report contained the following remarks.

This hospital is a small one but answers very well as a temporary arrangement. The ward is divided into two, the smaller part being used for convalescents as a day room. The walls have been rendered less bare by some very nice pictures given by Lady Hamilton ... A large Families' Hospital is greatly needed for the maternity cases.[8]

Although it was now known as the 'Military Hospital Tidworth', it continued to be shown on the Ordnance Survey maps as 'Delhi Military Hospital' until 1961[9] while on the 1976 edition[10] it was marked as 'Delhi Barracks (Military Hospital)'. In 1929 *The Barrack Book* still described the buildings as 'Delhi Barracks' and gave the 'usual occupation' as '1 Infantry Battn' and although it stated that the number of hospital beds were 'officers 7, men 230, families 24' it included a note which read '1 married and 824 unmarried men's accommodation misappropriated as temporary hospital'.[11] As the hospital had been in operation for 23 years, the word 'temporary' hardly seems appropriate. In course of time a new hospital was built on the site and in 1966 this consisted of eight wards: surgical (two), recovery, medical, officers', families', childrens' and maternity. There were also two operating theatres, an intensive care unit, a dental department, outpatients department, X-ray section and a laboratory.[12] When the hospital closed on 31st March 1977, the buildings were taken over by other units and in 1984 these included No. 2 Medical Group RAMC, Headquarters No. 1 Signal Group, Royal Corps of Signals and Headquarters Army Fire Services, United Kingdom Land Forces.

A number of other buildings of a standard design were built close to the main barracks but detached from them. These formed a series of groups each of which included the officers' mess and, a little distance away, the commanding officer's house which was named after the barracks to which it belonged, as for example Jellalabad House. In the case of Lucknow and Mooltan, a pair of semi-detached houses were provided for the respective commanding officers. These groups were built on the south side of Grand Trunk Road with the exception of Lucknow and Mooltan where they were situated on the east side of St. George's Road and to the north of Nainital Road. Between each pair of officers' messes were two semi-detached houses known as lodges, each being allocated to one of the barracks after which they were named, such as Bhurtpore Lodge. They were originally built for the quartermasters of the regiments that occupied the respective barracks.

The first officers' quarters were built in Park Road in 1907 opposite what is now The Oval and immediately to the south of Station Road.[13] These comprise a row of ten houses, the end house, next to Station Road, bearing the name plate 'Hampshire Cross', presumably in reference to the cross roads formed by Station Road, Lahore Road and Park Road. However there appears to be some difference of opinion as regards the location of Hampshire Cross, many old residents considering it to be at the junction of Bulford Road, Lahore Road and Meerut Road. By this time the barracks, the married quarters and most of the other military buildings had been completed, and the fact that 80 years later many of them were still in use and in very good condition is a tribute to those who built them. The contractors who were responsible for the construction of Tidworth Garrison were Henry Lovatt Ltd., Darlington Street, Wolverhampton whose London office was at North End House, Fitz George Avenue, Kensington.[14]

Although an Ordnance Survey map of 1903[15] marked the approximate position of the Royal Army Ordnance Depot as adjoining what is now Ordnance Road, no details were shown. However a plan which was first published in about 1918 and re-issued in 1931, shows that five buildings had been erected including the two tall stores which were still in existence in 1984.[16] Examination of this map shows that while the Ludgershall–Tidworth railway line had opened for goods traffic in 1902, neither a railway connection nor sidings in the Depot are shown. By 1923 considerable development had taken place: the number of buildings had more than doubled, sidings had been laid with the necessary connections while the site now covered almost 24 acres.[17] In December 1963 the Depot closed and since then the original area has been reduced by approximately one half although at the same time, the southern boundary has been extended. Part of the Depot was taken over by the Corps of Royal Electrical and Mechanical Engineers when the Corps was formed in 1942 and by the end of 1984 the Royal Army Ordnance Corps had finally vacated the premises which were at that time occupied by 3rd Field Workshops REME.

In 1910 Ordnance House was built at the northern end of St. George's Road as the official residence of the officer in charge of the Depot[18] but in 1984 it was in use as a Family Centre. Another house which was built in the same year was the White Bungalow standing at the junction of Park Road and Church Lane which is unusual in that it was constructed of timber; in 1929 it was allocated to the Assistant Director of Medical Services.[19] Four large detached houses were built in Arcot Road during 1912: Clive House, Havelock House, Muir House and Hastings House, the first two being the residences of the officers who commanded the Cavalry and Infantry Brigades, respectively.[20]

Opposite Jellalabad Barracks in Kirkee Road and adjoining the camp railway, was the Bakery which supplied the Garrison with bread; its close proximity to the line enabling truck loads of flour to be unloaded directly into the building. On baking days the appetizing smell of fresh bread could be enjoyed in different parts of Tidworth depending on the direction of the wind but by 1984 the Bakery had been closed and the building converted into riding stables now known as the Bakery Stables. Next to the Bakery and facing Kirkee Road was the Butchery which dealt with the allocation of meat supplies to the various units: the building was still standing in 1984. Another feature of the Garrison which dated from the early days was Tidworth Garrison Market which was also known as the Tin Market since it was built of corrugated iron. It stood on the north side of Bazaar Road opposite the junction with Lowa Road but it was closed during the 1950s and the site is now used as a car park for the Garrison Theatre. The Market comprised more than twenty stalls the tenants of which included two butchers, two grocers, a fishmonger, two greengrocers, a chemist, a tobacconist, two outfitters, a draper and a watchmaker while there were other stalls which sold furniture, shoes, hardware, cakes, sweets, bicycles and wireless sets. Administered by the Garrison Adjutant, the Market was open from 1000 hours until 1900 hours on Tuesdays, Thursdays, Fridays and Saturdays and was governed by a set of rules known as 'Regulations'. These dealt with such matters as hygiene, rents payable for stalls, the prohibition of smoking and dogs in the market and so on.

A matter which was of considerable interest to those who were engaged in electrical engineering was the decision to use electricity for lighting the barracks and married quarters. Although the War Office had already adopted this form of power in the barracks at Aldershot, it was still in the early stages of development. The scheme which was carried out by the consulting engineers, Messrs. Kennedy and Jenkin, included the building of a central power station at Tidworth with a sub-station at Bulford which was then a hutted camp.[21] The main generating station was built in 1913 in Candahar Barracks between Bazaar Road and the camp railway, from which a short siding ran into the power house. A little distance away, a row of four houses were built in Bazaar Road for those working at the generating station and in 1984 the house at the west end carried a stone inscribed 'G.R. 1913. Staff Quarters. Power Station'. This date also appears over the entrance to the power station building.

When the War Department bought the Tedworth Estate in 1897, one of the most urgent matters was to decide the future use of Tedworth House. In December 1898 it was agreed that Col. J.M. Barklie who was supervising the construction of the new barracks should be allowed to use the house as offices and for quarters for himself and his staff.[22] As the military importance of Salisbury Plain increased, steps were taken to deal with its administration and the Army List for June 1899 contains the first entry relating specifically to the Plain.[23] In October 1900 the Monthly List shows that Maj.-Gen. T.A. Cooke was commanding the troops on Salisbury Plain with his headquarters as 'Tidworth House, Andover'.[24] Although during 1903 the Lists contained a section headed 'Salisbury Plain', it was not until January 1904 that the title 'Salisbury Plain District' appeared with the postal address of 'Tidworth House, Andover'. This designation continued in use until Southern Command Headquarters moved from Salisbury to Tidworth during the spring of 1905 and presumably absorbed Salisbury Plain District. Command Headquarters remained at Tidworth for the next four years but had returned to Salisbury by October 1909. In 1905, Lieut.-Gen. Sir Ian Hamilton, KCB, DSO was appointed General Officer Commanding-in-Chief and Tedworth House became his official residence from April 1905 until May 1909.[25] He was succeeded by Gen. Sir Charles Douglas, GCB who also occupied Tedworth House as a private residence. Early in the First World War Tedworth House became an Officers' Club but it was later taken over to provide accommodation for nurses, the *Barrack Book* recording that in 1929 'the usual occupation' was 'nursing staff'.[26] During the Second World War it was used as a club for American soldiers when the United States Army was stationed in Tidworth. For some time after the War it was again used for nursing staff who worked at the Military Hospital but when the Hospital closed in 1977 it became an officers' mess. Since then it has continued to be used for this purpose and in 1984 it was officially known as The Headquarters 1 Infantry Brigade Officers' Mess.

At the end of the nineteenth century, before the War Department had begun to acquire land on the Plain, there were two churches in use at Tidworth – the ancient church of Holy Trinity in North Tidworth and the, then, recently built church of St. Mary's in South Tidworth. Holy Trinity is built in the Perpendicular Style which was adopted from about 1350 to 1530 AD and a brief account of it has been given by Pevsner.[27] St. Mary's which adjoins the Shipton Bellinger road, not far from Tedworth House, was built by Sir John Kelk, Bart. and the foundation stone was laid by Lady Kelk on 29th April 1879. The architect was John Johnson and the church has been described at some length by Pevsner and Lloyd.[28] In addition to these two churches, there was a mausoleum situated at the extreme west end of the wood known as 'The Plantation' which extends from Tedworth House towards Humber Lane. Thomas Assheton Smith and his wife were buried there but they were apparently re-interred at a later date in the burial ground of the Mortuary Chapel at the eastern end of Church Lane. The mausoleum has since been demolished but the Chapel in church Lane was still in existence in 1984. As from 1st September 1972, St. Mary's Church was declared redundant and the parishes of North and South Tidworth were united to form a 'new benefice and parish of Tidworth' of which Holy Trinity became the parish church.[29]

The provision of a Garrison Church was not apparently considered to be a matter of any great urgency and although most of the buildings had been completed by 1907, no action was taken to provide a church for those serving at Tidworth. The first person to draw attention to this was the Rev. W.S. Jaffray, SCF (Presbyterian) who wrote a long letter to the War Office on 14th July 1908, the opening paragraph of which read:

> I desire to bring before the Advisory Committee at their next meeting the position in which Presbyterians are placed with regards to Church accommodation at Tidworth. Most members of the Advisory Committee are probably aware that notwithstanding the presence of an Infantry Brigade of five battalions and one Cavalry Regiment plus details consisting of Staff, RE, RAMC, ASC and AOC, there is no such thing at Tidworth Camp as a Garrison Church. So far as I can gather there is no intention on the part of the Government to build one although the authorities are at present erecting what is generally believed will be a theatre or concert hall.[30]

The full title of the Committee was The Presbyterian Advisory Committee while the 'concert hall' refers to the Garrison Theatre which was then in course of construction. Two other matters which the Padre mentioned in his letter are of interest. The first was in respect of a suggestion that Church services should be held in a dining hall at Mooltan Barracks which would entail a march of a mile each way for some regiments.

> The men greatly dislike Parade Services because they are invariably preceded by a very long and strict inspection – the strictest inspection of the week. To march the men a mile to a dining hall for service will certainly intensify that dislike and lessen the chance of the men deriving any benefit therefrom.

The second matter emphasized the need for a church for the following reasons.

> ... the want of a proper church building has prevented our voluntary evening service from becoming the success it might be and should be. People anywhere will not turn out to worship in a cheerless whitewashed barrack room. We have nearly 1000 men and 100 families at Tidworth. If the work there is to be given a fair chance a church is a first necessity.[31]

This letter received immediate attention and the following reply was despatched two days later.

> The Secretary of State will see Lord Balfour of Burleigh and the other members of the Advisory Committee in his room at the House of Commons on Wednesday 22nd inst. at 4 p.m. The subject will be church accommodation at Aldershot and Tidworth and Lord Balfour is therefore anxious that you should be present if possible ... You may regard this letter as a War Office authority to attend, subject of course, to the General's concurrence.[32]

As a result of this meeting the Advisory Committee 'resolved to endeavour to provide an iron church at Tidworth', the War Office having 'placed £400 at the Committee's disposal'.[33] In August 1908 a quotation for £300 from Henry Lovatt, Ltd., the firm which had built the barracks, was forwarded to the War Office and this appears to have been accepted. Although the site had not been released by 1st November and despite difficulties encountered early in December regarding the provision of 'carpets, chairs and tables, holy', the church, dedicated to St. Andrew, opened for worship on Sunday, 24th January 1909.[34] The Rev. W.S. Jaffray was posted to London on 1st October 1910 but he had the satisfaction of knowing that through his efforts, Tidworth Garrison had been provided with its first church. It was closed in 1964 when part of the building which was formerly the Ian Hamilton Wesleyan Soldiers Home was converted into the Garrison Church of St. Andrew and St. Mark (Church of Scotland, Methodist and Free Churches). A photograph of the original church taken in 1984, when it was used by the Sea Cadets, will be found on plate 12.

On 28th April 1898 the Salisbury Diocesan Synod of the Church of England appointed a committee which was designated 'The Salisbury Diocesan Committee for the Welfare of Soldiers', but by 1900 the name had been altered to 'The Church of England Committee for the Welfare of Soldiers'. When the status of the Committee was changed to that of a permanent Diocesan Board in 1903, a further alteration was made to the name and it became 'The Church of England Diocesan Board for the Welfare of Soldiers'. In fact the word 'Diocesan' was never used in the title while in 1911 the words 'on Salisbury Plain' were added. During the First World War, Royal Naval and Commonwealth Forces were stationed on the Plain and so as to include these, the name was changed once again to 'The Church of England Board for the Welfare of the Imperial Forces within the Diocese of Salisbury'. When the Board was registered with the Charity Commission in 1967, the title finally became 'The Church of England Board for the Welfare of Her Majesty's Forces within the Diocese of Salisbury'. Although the name of the Board has changed on several occasions, its objects have always remained the same, namely 'the spiritual and moral welfare of the Soldiers', which in latter years has included sailors and airmen. In a scheme negotiated with the Charity Commissioners in 1985, this was extended to 'promoting the spiritual and moral welfare of and providing such other charitable benefits as the Managing Trustees from time to time decide for Her Majesty's Forces within the Diocese of Salisbury and their dependents'.[35]

When the Rev. F. F. S. Smithwick was posted to Tidworth in October 1907, he found that the position regarding Church of England services was very similar to that which had been described by the Rev. W. S. Jaffray. 'Our only church was the dining hall of one of the barracks that was disused at the time' but this was only large enough to accommodate a limited number of troops. Consequently, it was necessary to hold four parade services every Sunday although during the summer months open air services were held in Tidworth Park.[36] Proposals for a church had been put forward in 1907 but little progress was made until the Church of England Board for the Welfare of Soldiers appointed a committee in February 1910 designated the Tidworth Soldiers' Church Building Committee. Subsequently this was re-named the Church of England Garrison Church Tidworth Building Committee. The Chairman was Brigadier-General L. G. Drummond, CB, MVO who commanded 7th Infantry Brigade which was then stationed in Tidworth, but the enthusiasm of the Committee was largely due to the encouragement of the Bishop of Salisbury, Dr. Wordsworth. Very sadly he died before the church was finished. The Board's Annual Report for 1910 recorded that the plans for the church, drawn up by the architect, Douglas Hoyland of London, had been approved by the Bishop and the Army Council. The estimated cost was 'about £4,800' and it was designed to seat 1,200 people.[37] It was suggested at first that the church should be 'a temporary corrugated iron building' but it was subsequently decided to build a permanent church of hollow buff coloured terracotta blocks.[38] Subsequently a tender of £4,946 submitted by Messrs. C. Grace & Sons of Ludgershall which included the architect's fees, was accepted[39] and on 22nd May 1911 the foundation stone was laid by Maj.-Gen. Sir Henry Rawlinson, Bt., CVO, CB, GOC 3rd Division.[40]

Within twelve months the Church had been completed and on 4th May 1912, the Chaplain-General dedicated it to St. Michael in the presence of Lord Haldane, the Secretary of State for War, the Bishop of Salisbury and 'all the Chief Officers of the Plain'. The final cost of the building was £4,639; the architect's fees, road making, furnishing, etc., added another £702.[41] During the past seventy years a number of alterations and improvements have been carried out, especially when the Sanctuary was enlarged and in 1984 it was only possible to seat 800.[42]

Unfortunately little information is available as regards the history of the Roman Catholic Garrison Church of St. Patrick and St. George and enquiries have proved to be unproductive. In an entry of a few lines, Pevsner and Lloyd state that it was built in 1912 and that the architect was G. L. W. Blount.[43] The Garrison Church of St. Andrew and St. Mark (Church of Scotland, Methodist and Free Churches) which was part of the original Ian Hamilton Soldiers' Home, was dedicated on 18th October 1964,[44] when the old Presbyterian Church of St. Andrew was closed. However, in September 1985 the Church of St. Andrew and St. Mark was also closed and the services transferred to St. Michael's where an inter-denominational service was held once a month.

With the completion of the barracks, units began to be stationed at Tidworth and it soon became evident that there was a very real need to provide for the welfare of soldiers when they were not on duty. The first to meet this challenge was the Wesleyan Church through its Army and Navy Board which by 1907 had begun to build a Wesleyan Soldiers' Home on a site at the junction of Kohat Road and Bulford Road. The estimated cost was £6,000 of which the Board contributed £3,000 and launched an appeal to raise the balance. The brochure issued in connection with this referred to the need to provide 'for the soldier's social and religious comfort and welfare'.

> Singularly enough, however, this important duty has hitherto been to a great extent overlooked; for whilst otherwise the Barracks are admirably appointed, neither Church, nor School building, nor Soldiers Home has yet been provided ... The urgent need for the immediate provision of this Home will be obvious from the following fact:– No Home or Institute exists in the barracks, the barracks are just completed but the permanent Force already numbers 4,000 men, to which very shortly a Cavalry Brigade will also be added, and during the Training Season, between April and September, 20,000 men encamp in the immediate neighbourhood.[45]

The new building included four large rooms: a lecture hall, concert and service room, a reading and writing room, a refreshment room and a billiard and games room together with the usual domestic premises.[46] It was built by Messrs. C. Grace & Sons and was opened on 4th July 1908 when it was named the Ian Hamilton Wesleyan Soldiers' Home after Lieut.-Gen. Sir Ian Hamilton, KCB, DSO who was then Commander-in-Chief. The visitors on the opening day included the Secretary of State for War, Mr. R.B. Haldane (afterwards Viscount Haldane of Cloan), Sir Ian and Lady Hamilton and Brig.-Gen. Sir Hubert Hamilton, CB, DSO commanding the 7 Infantry Brigade at Tidworth.[47] The lecture hall and service room at the west end of the building was dedicated as the Garrison Church of St. Andrew and St. Mark on 18th October 1964 and reference has already been made to this.

On 6th March 1908 the Church of England Board for the Welfare of Soldiers appointed a Tidworth Institute Management Committee. The purpose of the Committee was to provide a Church of England Institute for soldiers at Tidworth, and a site in Kirkee Road between the Presbyterian Church and the Bakery was selected. The Institute which was opened by Gen. Sir Charles Douglas, KCB and the Bishop of Salisbury on 27th August 1910, had previously been visited by FM.

HRH the Duke of Connaught, KG.[48] The building adjoined the Electric Cinema but was demolished in about 1965. A Roman Catholic Institute was built on the south side of St. Patrick's Avenue and to the south-west of the Roman Catholic Church. It has not been possible to obtain any information relating to this institute except that it was closed in 1964 and demolished shortly afterwards.

The institute which has the longest record of service at Tidworth, is the Red Shield Services Club administered by the Salvation Army. This occupies a site in Baroda Road to the north of Clarendon Terrace and the fact that it was opened in October 1914 and was still in operation in 1985 – seventy years later – is a very fine achievement.[49] Early in the appalling winter of 1914, action was also taken by the Young Mens' Christian Association to provide soldiers with somewhere to relax and enjoy themselves when they were off duty. More than twenty wooden huts were erected on the Plain and run by the Association, four of which were at Tidworth. These were sited in Aliwal, Bhurtpore, Candahar and Lucknow Barracks, but during the 1930s they were demolished and replaced by a permanent brick building at the junction of Bazaar Road and Dasna Road.[50] This continued to serve those stationed at Tidworth throughout the Second World War but in view of the changing conditions in Army life, it was closed in 1974 and demolished at the end of 1978.[51] In addition there were also a number of canteens run by the Navy and Army Canteen Board which in January 1921 became the Navy, Army and Air Force Institutes more usually referred to by the initials NAAFI.

The War Office also realized that with the large numbers of troops stationed in Tidworth, something should be done to provide facilities for entertaining them. Consequently on 28th February 1908 work began on the construction of the Garrison Theatre at the junction of Bazaar Road and Lowa Road. The work was completed on 19th March 1909 and on 4th June the theatre opened with a boxing competition held by the Royal Fusiliers.[52] Mr. Bert Pickernell was appointed stage manager and then manager, and continued to hold this post until his retirement in 1964 when he was succeeded by his son, Ken, who had just left the Royal Air Force. He was still managing the Garrison Theatre in 1985. Since 1909 the Theatre has been the scene of many events: orchestral and military band concerts, variety shows, regimental dinners, boxing tournaments, film shows and lectures. Amongst those who have appeared in the Garrison Theatre are Dame Nellie Melba, the Australian opera star in 1926, Sir Adrian Boult, Alistair Sim, Will Hay, Norman Wisdom, Eric Sykes, Henry Cooper and Frankie Vaughan. At the beginning of the Second World War the theatre was taken over by the Entertainments National Services Association better known as ENSA. Later when units of the United States Army were stationed in Tidworth from 1942 to 1944, the entertainers included Bob Hope and James Cagney. During the Second War, the lecturers on military matters included FM Earl Alexander of Tunis and FM Viscount Montgomery of Alamein.[53]

Three cinemas have been built in Tidworth of which the first was the Electric Cinema which adjoined the Church of England Institute in Kirkee Road and was sometimes referred to as the 'Institute Cinema'. The Institute itself was built in 1910 but as electricity was not available in Tidworth until 1913 when the power station was built, the name would suggest that the cinema was not in operation until that year, or shortly afterwards. It closed in the early 1960s and the seats were removed but it was still in use as a recreational building in 1985. The Hippodrome Cinema which seated 612 people was probably built during the First World War since it is not shown on an Ordnance Survey Map of 1914 but is marked on one of 1926 which had been revised in 1923. It was situated in Pennings Road between what is now Service Garage and the Royal British Legion Club. During the night of 11th November 1963 flames were seen issuing from the roof and despite the efforts of six fire brigades it was completely destroyed.[54] A new cinema, the Globe, was built in 1967 to the north of Station Road opposite the new NAAFI buildings which occupy the site of the old Tidworth railway station.

The first school in Tidworth was built by Mrs. Assheton Smith in 1857[55] and although in a derelict condition it was still standing at the junction of Station Road and Pennings Road in 1985. Originally known as 'Mrs. Assheton Smith's School', by 1880 it had become the 'North and South Tidworth National School' and in later years the Tidworth Church of England Primary School. It was closed in June 1980 and the children were transferred to the 'Army School' in Bazaar Road opposite the Garrison Theatre which was then renamed the Tidworth Church of England (Controlled) Primary School. This had been built to provide education for the children of soldiers in Tidworth Garrison and the principal building which is of brick and adjoins Bazaar Road, carries a stone inscribed 'ER 1909' recording the date of its erection. On the other side of the playground is a range of classrooms constructed of timber which would appear to have been erected during the First World War or shortly afterwards, as it is shown on an Ordnance Survey Map of 1926 (1923 revision) but not on earlier maps. On 1st April 1948 the school was taken over by the Local Education Authority from the Army.

During the last century, Army education followed an uncertain course which was subject to continual changes in policy and organization but one factor which provided a degree of stability was the formation of the Corps of Army Schoolmasters on 2nd July 1846.[56] In 1857 a Council of Military Education was set up which subsequently took over the control of Army Schools and Libraries[57] while in 1870 a Director-General of Military Education was appointed.[58] This post was terminated in 1889 and was replaced, three years later, by a Director of Schools. In 1920 the Corps of Army Schoolmasters was replaced by the Army Educational Corps which became the Royal Army Educational Corps in 1946. Since 1840 much of the teaching in Army childrens' schools had been carried out by Schoolmistresses and in recognition of their work they were given the designation of 'Queen's Army Schoolmistresses' in 1927.[59]

When the proposed acquisition of land on the Plain was announced, the Secretary of State said in the Commons that 'there was no intention of erecting a great number of permanent barracks on Salisbury Plain' and that 'the object ... was to afford a manoeuvring ground for the troops, some of whom would probably belong to the Militia or the Volunteers'.[60] Although the decision not to erect permanent barracks was later rescinded there was, from the start, a need to provide camp sites for troops during the training season. One of the first references to these occurs in the minutes of a meeting of the Salisbury Plain Committee held on 18th November 1897, when note was taken of a report by 'Southern District' concerning 'ten camping grounds four of which would be standing camps'.[61] At a later meeting, Camp No. 9 which appeared to be Park House and No. 10 which was Bulford, were discussed.[62]

Five camps at or near Tidworth were shown on a map published by the Ordnance Survey in 1903.[63] These were:

Park House, three quarters of a mile south-west of Shipton Bellinger;
Perham Down, on the site of the present barracks;
Tidworth Park, an L-shaped site covering a rectangular area between the Shipton road and Ashdown Copse which extended westwards, south of the present sewage works;
Tidworth Pennings, between Sidbury Hill and the Military Cemetery, and
Windmill Hill, three quarters of a mile north-west of Ludgershall.[63]

Although all these camps were used, at one time or another, by Regular and Territorial units, Tidworth Park and Tidworth Pennings were usually allocated to Officer Training Corps for the month of August between the two world wars. A list of camps on Salisbury Plain will be found in Appendix IV.

By the end of 1912, except for the power station, the buildings had been completed, the Garrison was well established and its life in full swing. With the British Empire as yet unchallenged, it was a period of intense national pride in which the Forces of the Crown played a full part. Although the soldiers of the Queen were now soldiers of the King, the same *esprit de corps* existed and pride in the Regiment was paramount. The Army had also found a champion in Rudyard Kipling whose poems and prose showed the British soldier in a new perspective. On a more practical note the Army had gained a great deal of experience in the Boer War and this, in conjunction with Haldane's intro-

5.ORIGINAL ARMY CAMPS NEAR TIDWORTH

(As referred to in Appendix IV)

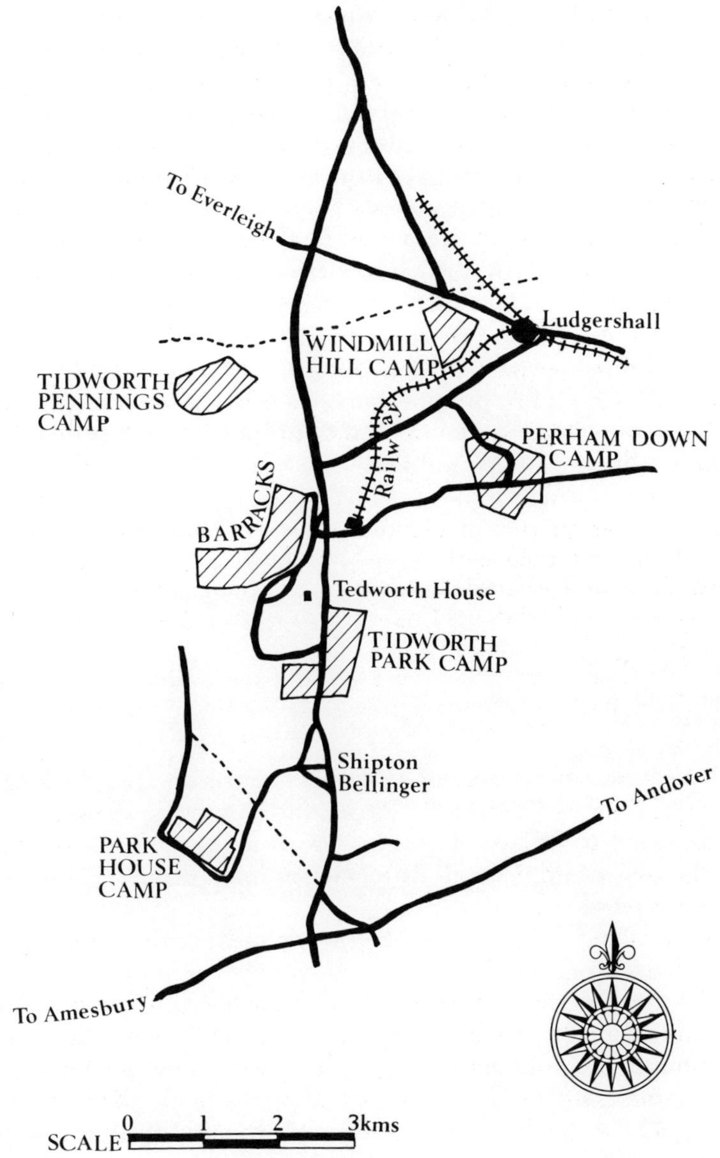

ductions and reforms of 1908, were creating a new professional approach to military matters.

Although khaki had been introduced during the War in South Africa, regiments continued to wear their traditional uniforms in peacetime, on occasions such as church parades, guard mounting, reviews and inspections. However the greatest single factor that contributed to the unique quality of army life before the First World War, was the horse. Apart from the fact that the Army depended on horses and, to a lesser extent, on mules for its transport, there were three cavalry barracks in Tidworth. Consequently horses were present everywhere: on parade, on exercise, drawing General Service wagons, limbers, ambulances, field kitchens and so on. They also provided an audible background with the jingle of bits, the grind of iron shod wheels on hard roads and, above all, the unmistakable sound of hooves striking the ground.

Today standing at the bottom of St. Michael's Avenue and looking down Grand Trunk Road one can still imagine the scene on a Sunday morning, seventy years ago. At first, the distant sound of a band as a regiment left Aliwal Barracks and began its march to the Garrison Church. When it had passed Assaye it would be followed by the next unit with its band and so on until each succeeding regiment was on its way, while those from Lucknow and Moultan would approach from the opposite direction. It is unlikely that many of those who watched these regiments marching past, realized that they were witnessing the end of an era. August 1914 was not far away.

CHAPTER SIX

Tidworth – War and Peace

By 1914 Tidworth was second only to Aldershot as a military centre while its proximity to what was now the finest training area in the country, added considerably to its importance. In the meantime the Garrison had become a compact self-contained community with its own shops, churches, schools and railway station. In August 1914 two brigades were quartered in Tidworth – the 2nd Cavalry Brigade and the 7th Infantry Brigade. The former, commanded by Brig.-Gen. H. de B. De Lisle, CB, DSO, was composed of the 4th Royal Irish Dragoon Guards, the 9th Queen's Royal Lancers, the 18th Royal Hussars and the 2nd Signal Troop, Royal Engineers. The Infantry Brigade under the command of Brig.-Gen. F.W.N. McCracken, CB, DSO comprised the 3rd Bn. The Worcestershire Regiment, the 2nd Bn. The South Lancashire Regiment, the 1st Bn. The Wiltshire Regiment and the 2nd Bn. The Royal Irish Rifles. There were in addition the 20th Company Royal Army Medical Corps and a section of the 2nd Company Army Ordnance Corps.[1]

Although the situation in Europe had been deteriorating for some time, this does not appear to have been appreciated by the general public or indeed, by some who ought to have been aware of the trend of events. On 8th July 1914, Gen. Sir Horace Smith-Dorrien remarked that

> there were few people in England who realised the gravity of the situation. We soldiers had had so many firsthand examples of how Germany was preparing, and had failed to convince those in high places, that I was almost in despair.[2]

However, when the storm finally broke, the Army reacted immediately both on the staff and in individual regiments.

> Andover Flower Show was held every year on August Bank Holiday Monday in the Walled Meadow. The Cavalry Brigade from Tidworth always put on a display of tent-pegging, etc. On 4th August 1914, news of the outbreak of war reached them as they were preparing to start their display. The Lancers and Dragoon Guards were taking part. They hitched up their horses to their GS wagons and went back to Tidworth at full gallop, without giving their display.[3]

With the declaration of war came a flood of volunteers who were followed by those of Lord Kitchener's New Army, and in anticipation of the inevitable concentration of troops at the chief military centres, steps were taken to provide huts for living accommodation and so on. Orders for these were given by the War Office on 12th August and within two days complete working drawings for a hutted camp, large enough to accommodate an infantry battalion at war strength had been prepared and approved by the Army Council. The drawings consisted of seventeen different designs which covered accommodation huts, cookhouses, lavatories, officers' and sergeants' messes, officers' quarters and other related buildings. This remarkable achievement was due to the efforts of Maj. B.H.O. Armstrong, RE, his chief assistant Mr. J.D. Michel and their staff of draughtsmen who worked continuously for 48 hours.[4] Subsequently Maj. Armstrong designed a sectional hut which could be erected with local labour and was known as the 'Armstrong Hut'.[5] On the Plain, hutted camps were built at Tidworth and Bulford but the largest was at Larkhill which was described as 'a new hutted town'.[6]

Glimpses of life in one of these hutted camps in 1916 has been recorded by members of the 2/5th Bn. The Gloucestershire Regiment.

> On the night of February 19th, the Battalion reached Tidworth and marched to its quarters at Parkhouse Camp. First impressions of life on the Plain were not very encouraging: the weather was bitterly cold; there were no palliasses; there were no fires and no light as the electricity failed on the night of the Battalion's arrival. The men were put into huts, given rations and three blankets apiece and left to get what sleep they could on bare bed boards.[7]

During the next three months the Battalion dug trenches, went on frequent route marches, had a great deal of rifle practice and took part in numerous field days and exercises in the area around Silk Hill, Quarley, Cholderton and Andover.

On 5th May the whole Division was inspected by HM King George V, a sure sign of their impending departure overseas.

> We marched off at 9 a.m. It was close and muggy, but rain during the previous night had laid the dust. As the inspection had been well organised, it was not so long drawn out as most operations of a similar nature. It was a most impressive sight. Twenty thousand men, hundreds of horses and wagons all moving together in an endless line, under the command of a single man – all very fine, fixed bayonets, drawn swords and bands playing.

> One does not wish to be a pessimist, but I could not help wondering how many would come back alive from France.[8]

On 24th May 1916, they entrained at Tidworth Station for Southampton and France.

> The Battalion had arrived on the Plain under the worst possible conditions, but despite the discomforts and the tedium of training, Parkhouse Camp had quickly become its home, and though doubtless no one was sorry for a change of venue, yet when the day of departure arrived, there was a feeling of reluctance to leave the Plain and the memories it held in its undulating folds.[9]

In the middle of October 1914, Canadian and Newfoundland troops landed at Plymouth and then moved to camps on West Down between Shrewton and Urchfont and these are described in Chapter 10, while in January 1915 a Canadian Training Depot was set up in Tidworth.[10] When the Australian Imperial Forces proceeded to France after the Gallipoli campaign, it was decided to form Australian training battalions in England in order to provide reinforcements for their Divisions at the Front.[11] These were divided into three groups numbered with reference to the Divisions to which the reinforcements would be posted and located as follows.

No. 1 Group for the 1st Division, at Perham Down
No. 2 Group for the 2nd and 3rd Divisions, at Rollestone
No. 5 Group for the 5th Division, at Larkhill.

Other arms and corps were allocated depots in the following locations

Artillery Depot	Larkhill
Engineers and Signallers Depot	Parkhouse
Army Service Corps Depot	Parkhouse
Army Medical Corps Depot	Parkhouse
Machine Gun Corps Depot	Tidworth Barracks.[12]

Personnel for these training units began to reach the United Kingdom early in June 1916[13] but very soon a reorganization of the depots and their localities took place, the engineers, signallers and machine gunners moving to other parts of the country.[14]

By August 1917 the Headquarters of the Australian Imperial Force Depots in the United Kingdom had been established in Bhurtpore Barracks and were still there in March 1919 while between 1916 and 1917 part of Candahar Barracks were occupied by other units of the Australian Army.[15] In July 1916 the 3rd Australian Division arrived in England from Australia and was stationed at Larkhill until it left for France at the end of 1916.[16] Four 'Command Depots' were also set up to deal with men who had been discharged from hospital and of these, No. 1 was located at Perham Down. Each one was concerned with a different hospital category of men and Perham handled those who were considered to be fit enough to return to their units in France.[17] At the end of the war, large numbers of Australian troops returned to England and were quartered at Bulford. The last 10,000 men were brought back from France in May 1919 and joined the 70,000 who were already on Salisbury Plain awaiting their return home.[18] Australian troops were also stationed at Codford and Sutton Veny and information relating to these camps will be found in Chapter 11.

In March 1916 there were approximately 2,000 convalescent New Zealand soldiers in England who had been evacuated from Gallipoli and who, when they were passed fit, were posted to the New Zealand Army Depot at Hornchurch.[19] However, by the beginning of June a 'command depot' for recuperating personnel had been established at Codford with a training depot at Sling Camp, Bulford,[20] and further reference to these will be found in Chapters 11 and 8 respectively. One hundred New Zealand and one hundred and fifty-nine Australian soldiers who died during the First World War in Tidworth and Fargo Hospitals are buried in Tidworth Cemetery.

After the War, Tidworth returned to peacetime conditions and, to some extent, to its pre-war way of life but changes were not far distant, the most significant being the increase in mechanization which in course of time, was to lead to the elimination of the horse for all practical purposes. In January 1922, as in August 1914, the 2nd Cavalry Brigade and the 7th Infantry Brigade were stationed at Tidworth. The Monthly Army List shows that the Cavalry Brigade was composed of the 19th Royal Hussars, the 20th Hussars, the 5th Royal Irish Lancers, the 17th Lancers and the 21st Lancers, detachments of the 4th Royal Irish Dragoon Guards and the 9th Queen's Royal Lancers and the 2nd Signal Troop Royal Corps of Signals. It would seem that the 7th Infantry Brigade had not yet been re-formed as the Army List does not show any complete battalions but only detachments of the 1st Bn. The Gloucestershire Regiment and the 2nd Bn. The Loyal Regiment.[21] On 11th April 1922, as a result of amalgamations, eighteen cavalry regiments were reduced to nine and in January 1924 the composition of the two brigades at Tidworth was as follows. The 12th Royal Lancers, the 14th/20th Hussars, the 15th/19th Hussars, the 17th/21st Lancers and the 2nd Signal Troop, Royal Corps of Signals formed the 2nd Cavalry Brigade while the 1st Bn. The Lancashire Fusiliers, the 2nd Bn. The Loyal Regiment, the 2nd Bn. The King's Shropshire Light Infantry and the 1st Bn. The Wiltshire Regiment formed the 7th Infantry Brigade.[22]

Although, as already mentioned in the last chapter, the Army Lists for 1903 included a section headed 'Salisbury Plain', this was changed in January 1904 to 'Salisbury Plain District' but after the spring of 1905 there was no further reference to it. However, the Army List for July 1920 contained a section entitled 'Salisbury Plain Area' and this was defined as 'comprising the County of Wiltshire and Tidworth' such qualification being necessary since part of Tidworth lies in Hampshire, the Area Commander being 'The GOC 3rd Division, Bulford'. In January 1932 the boundaries were extended to include 'all War Department land on or about Salisbury Plain and Salisbury, Trowbridge and Chiseldon'.[23] Seven years later, this definition was replaced by a very detailed description of the area,[24] while in April 1941 it was shown as being divided into garrisons and sub-areas. In January 1942 the title was again changed to 'Salisbury Plain District' which consisted of the Garrisons of Tidworth, Larkhill, Warminster, Devizes and Chiseldon together with the Sub-areas of 'North Wilts' and 'South Wilts'.[25] However there was no reference to Salisbury Plain District in the Army List in the following April or in any subsequent issues.

In 1920 an event took place which, during the course of the next 18 years, was to bring Tidworth to the notice of many thousands of people who might otherwise have hardly been aware of its existence. This was the first performance of the Tidworth Tattoo which owed its origin to Maj.-Gen. T. T. Pitman, CB, CMG who from 1920 to 1922 commanded the 2nd Cavalry Brigade. He had already staged a successful tattoo in 1919 when serving with the British Army on the Rhine but at Tidworth his main object was to entertain members of the Garrison, their families and those who lived in the neighbourhood. The first tattoo took place on 8th, 9th and 10th June 1920 on the polo ground in front of Tedworth House, with the searchlights installed on the roof of the mansion. No seating arrangements were provided other than a limited number of seats for officers, but this did not prevent it from being an outstanding success. Gen. Pitman had decided that any profit should be given to military charities and for the first year this amounted to about £400 which by 1920 values was a very satisfactory result.[26]

In the following year difficulties occurred due to a shortage of army personnel brought about by service requirements in Ireland and duties in connection with a number of strikes. As a result, Royal Navy reservists were moved to Tidworth to replace some of the troops and these took part in the Tattoo that was held on 26th and 27th May 1921. The last Tattoo which was organized by Gen. Pitman was that of 1922 and in the following year Col. Comdt. G.V. Hordern, CB, CMG who was commanding 7th Infantry Brigade, became President of the Tattoo Committee. At the same time, Lieut.-Col. C. de Putron who commanded the 1st Bn. The Lancashire Fusiliers, was appointed Chairman and became responsible for drawing up the programme and organizing the rehearsals while Capt. J. C. Latter of the same battalion was made Honorary Secretary. A temporary open grand stand of about 200 seats, was erected at the southern end of Tedworth House gardens as well as other seating facilities and the receipts rose to £1,100.[27]

When Brig. B.D. Fisher, CB, CMG, DSO who commanded 2nd Cavalry Brigade, took over the organization of the Tattoos which were held from 1924 to 1926, he decided to allocate any profits towards improving the facilities for games and other amenities in Tidworth

Garrison. Many of the sports grounds which are in use today were laid out with the help of these grants and since the net profit in 1926 was £4,044 this financial assistance was well worth having. By 1925 it was clear that with the increasing number of spectators, a larger ground would soon be needed which could also be used for horse shows, gymkhanas and other similar events and consequently Brig. Fisher decided to construct a new arena. In order to do this it was not only necessary to level the site but also to form raised areas on which stands could be erected. This involved the moving of over 10,000 tons of soil and the laying of about 100,000 squares of turf which were obtained from Perham Down. Except for some assistance in laying the turf, the whole of the work was carried out by troops in the Garrison. This formidable task began in January 1926 and was finished by the middle of April ready for the Tattoo which was held in July. In order to provide reserved seats for 7,000 spectators, removable timber terraces were constructed, on which chairs and benches were placed and these were stored in the Royal Army Ordnance Depot when they were not in use.[28]

In 1926 Brig. Fisher left Tidworth and in the following year the management of the Tattoo was taken over by Southern Command and it was re-named 'The Southern Command Tattoo'. No sooner was the 1926 Tattoo over, than Brig. Fisher initiated the extension of the arena and with the completion of the work in the spring of 1927, it covered an area which measured approximately 325yds by 160yds (297m. × 146m.). As the attendances rose, difficulties were experienced in controlling the crowds largely on account of the dilapidated state of the fences around the site. To deal with this, some 563yds (515m.) of fencing were erected in 1927 with further additions in the following year, as well as nine turnstiles. At the same time extra timber terracing was constructed and since the removal of this, at the end of each summer, was becoming a major operation, it was decided to leave it in position permanently. The existing terracing was also extended and increased in height so as to provide additional seating while a new stand was erected in the south-west corner of the arena. However the most important improvement was carried out in 1929 when a covered stand, 132ft. long, was built which provided seating for 1,025 spectators. This included a Royal Box, a control platform constructed in the roof, lavatories and storage facilities, the estimated cost being just under £3,000. As a result of these improvements the total seating accommodation was increased to approximately 13,750 with an additional standing room for 10,000.[29]

To meet the increasing demand, a Tattoo office was opened in 1928, in Kirkee Road next to the Bakery which was opposite Jellalabad Barracks and at the same time a paid secretary was appointed. After 1926 when Southern Command Headquarters took over the management of the Tattoo, Col. W.J. Dugan, CB, CMG, DSO became responsible for the actual production from 1927 to 1930.[30] During the

1930s, the Tattoo provided an outstanding military display which was extremely popular with the general public, and improvements were continually being made throughout this period. Each tattoo followed a fairly regular pattern and a summary of the 1935 programme will be found in Appendix V. During that year over 2,000 performers took part as well as many other Army personnel who were concerned with the organization behind the scenes. Five large searchlights, each of 40 million candle-power were sited on the flanks of the arena with an inner line of seventeen smaller ones having a power of thirty million each, making a total of 710 million candle-power. As regards communications more than sixty-five miles of telephone wires were laid with an exchange in the control tower. The whole cost of the Tattoo was met out of the receipts and the balance was dispensed as grants to welfare and recreational funds.[31] Those taking part saw the Tattoo in a very different light to the spectators. Looking across the darkened arena between each event when the searchlights were turned off, the stands appeared as a mass of little flickering lights as hundreds of cigarettes and pipes were being lit – 'it was all that you could see'.[32] As soon as the 1938 Tattoo was over, work was put in hand for 1939 but in view of the international situation it was decided on 15 May that the Tattoo for that year would have to be cancelled.

Tidworth did not produce another tattoo for twenty-seven years and when it did it was only a shadow of its former self, only occupying one afternoon's daylight performance instead of the seven floodlit evenings of 1935. It was held at 2.0 p.m. on Sunday 22nd May 1966 following the Army Three Day Horse Trials and in referring to the Trials *The Andover Advertiser* commented

> An innovation that will add to the usual thrills on Sunday afternoon is a military tattoo – the first at Tidworth since World War II.[33]

It was announced that the programme would include massed bands, physical training display, free fall parachuting by the Red Devils, a mock battle by the Parachute Squadron of the Royal Armoured Corps and a helicopter display. Unfortunately it was not possible to complete the programme as *The Advertiser* reported,

> The famous pre-war Tidworth Tattoo arena came back to life again on Sunday afternoon when a military tattoo was presented to wind-up the Army's three day horse trials. Upwards of 8,000 watched the colourful display which was unfortunately marred by a high wind. It made it difficult for people to hear the bands properly but it also meant the cancellation of one of the highlights of the afternoon's entertainment – free fall parachute dropping by members of the Parachute Regiment. Not only was the wind too high to make such daring drops but it was too strong for the aircraft to take off from Netheravon with the parachutists.[34]

However, even in its modified form, the Tattoo was only to last for another ten years and the last occasion on which it was held was 31st

May 1976. So came to an end a series of outstanding displays which, between the wars, had reached a standard of excellence which could hardly be surpassed.

An event of purely domestic interest took place on 28th January 1920 when the War Department sold The Ram Inn in North Tidworth to the Portsmouth United Breweries Ltd. (now part of Whitbread & Co. Plc.) for £7,000. It had formed part of the original Tedworth Estate that the War Department had bought from Sir John William Kelk in 1897 but since 1920 the Inn has been completely re-built. Standing at the junction of Pennings Road and Ludgershall Road, it has become a landmark and an institution where, for more than eighty years, many generations of soldiers have enjoyed a drink.[35]

In an attempt to improve the unsatisfactory arrangements regarding army canteens, three serving officers set up the Canteen and Mess Co-operative Society in 1894 and it became one of more than 150 caterers approved by the War Office. After the outbreak of war in 1914, there was an urgent need to provide canteens for the British Expeditionary Force in France and the Canteen and Mess Society, in conjunction with Messrs. Richard Dickson & Co., were asked to form an Expeditionary Force Canteen Organisation which came into operation in France in February 1915. However, in order to overcome certain difficulties that had arisen in connection with the financial aspects of army catering, the Army Canteen Committee was set up on 1st January 1917 as a 'company not for profit'. In the following June, the Royal Navy asked if it could be included and the Committee became The Army and Navy Canteen Board. The Expeditionary Force Canteens continued to function from 1918 until December 1920 but in January 1921 a body known as The Navy, Army and Air Force Institutes was formed, and took over the Navy and Army Canteen Board. Generally known by the initials 'NAAFI', establishments were set up during the 1920s at Tidworth, Bulford and Larkhill, the area administrative offices being at 47, Endless Street, Salisbury. During the summer, canteen facilities were provided at Officer Training Corps and Territorial tented camps where the initials 'NAAFI' were synonymous with some of the more enjoyable but less expensive things in life.[36]

As those who had served in the Forces returned to civilian life after the end of the First World War, various proposals were put forward for the formation of an ex-service men's association and subsequently the British Legion (now the Royal British Legion) was established in 1921. A building for the Tidworth Branch which was formed on 12th July 1922, was erected shortly afterwards on the east side of Pennings Road almost midway between Station Road and Ordnance Road.[37] It was replaced by the existing building which occupies approximately the same site and was opened by HRH The Duke of Gloucester, KG, KP, KT, GCB, GCMG, GCVO on 27th October 1962. Close to the British Legion building was The Bungalow Hotel which was built in

1922 by A.N. Kendal in Pennings Road about eighty yards to the north of Station Road. The object was to provide accommodation for single officers but in the early 1930s it was converted into flats and renamed The Albany: it was demolished in 1982.[38] A.N. Kendal also built the Hippodrome Cinema at Tidworth which was burnt down in November 1963 and a cinema in Bulford which stood at the junction of Marlborough Road and Rawlinson Road. In order to discourage pedestrians from walking across it, the Oval was enclosed by a boarded fence during the summer of 1924.[39]

In 1922 those regiments which were stationed in Tidworth began to build their own sports grounds subsequently helped, no doubt, by the Tattoo profits which Brig. B.D. Fisher made available for this purpose in 1924. Some idea of the results which were ultimately achieved, can be obtained by comparing successive editions of the Ordnance Survey maps of the Garrison area. The 1926 edition (revised in 1923)[40] shows eight grounds which included the Oval, the cricket ground on the Shipton road and the recreation ground adjoining Aliwal Barracks while the edition of 1936 shows that these had increased to thirteen.[41]

Tidworth has always been noted for polo and the first steps to provide a polo ground have been recorded in the minutes of a meeting of the Salisbury Plain Committee held on 18th May 1898.

> Polo Ground, Tedworth.
> The Chairman notified that the Inspector-General of Cavalry had raised the question as to the possibility of having a polo ground on the Tedworth estate for the use of the Cavalry Division. A reply had been given that the ground proposed by him was already let, but that there would be no objection to its use as a polo ground, provided that he settled with the tenant and that no expense to the War Department was involved.[42]

Unfortunately the location of the intended ground was not recorded and it is probable that in the early days polo was played on a suitable field rather than on a ground that had been specially laid out. Some support for this view is lent by the fact that on none of the War Department maps which were published before the First World War (the last being revised in 1914) is a polo ground marked. However an Ordnance map which is understood to have been published in about 1918 although later revised,[43] shows a polo ground on the east side of the Tidworth–Shipton road (A338) and opposite St. Mary's Church. This became known as 'No. 1 Ground'.[44] The next Ordnance map of the area to be published, was issued in 1924 – it had been revised in 1923 – and this showed two grounds: No. 1 Ground, and another at the eastern end of the Park immediately to the south of Tedworth House and extending from Arcot Road to the River Bourne. It was on this ground that the first Tattoo took place in 1920.

> The Tattoo and the succeeding Tattoos, until the year 1925 (inclusive), were held on what was at that time the polo ground situated in front of Tidworth House.[45]

As mentioned earlier in this chapter, Brig. B. D. Fisher who commanded the 2nd Cavalry Brigade, took over the organization of the Tattoos from 1924 until he left Tidworth in 1926. In 1925 it was decided that a new arena should be laid out and that it should be large enough to be used for recreational purposes when not required for the Tattoo. It was also agreed that part of the area should be used as a polo ground and it subsequently became known as the 'Fisher Ground' after Brig. Fisher. It is said that most of the work was carried out by the 2nd Cavalry Brigade which comprised the 12th Lancers, the 14th/20th Hussars, the 15th/19th Hussars and the 17th/21st Lancers and that the fatigue parties included all the officers from the Commanding Officers downwards armed with picks, shovels and wheelbarrows.[46] The Fisher Ground was considered to be second only to Hurlingham and was reserved for the Inter-Regimental Tournament.[47]

During the 1930s polo became so popular at Tidworth that steps were taken to provide additional grounds. The first was laid out at Tidworth Pennings on the west side of the Collingbourne Ducis road (A338) while at Perham Down the small ground which had been in existence since before 1923, was doubled in size. In Tidworth Park, the 'Tattoo Ground' which lay to the west of the Fisher Ground and in front of the Tattoo stand, was also brought into use.[48] No. 1 Ground which had not been used for polo since 1939 was taken over as a Rugby football ground in 1945[49] and was still being used for this purpose in 1985. Of the numerous visits which were paid to Tidworth during this period, one of the most notable was that of HM King Alfonso of Spain, KG, GCVO on 8th July 1926. After inspecting the 16th/5th Lancers (now the 16th/5th The Queen's Royal Lancers) of which he was Colonel in Chief, His Majesty played polo for the Regiment during the afternoon when it defeated the 7th Queen's Own Hussars (now The Queen's Own Hussars).[50]

Life in the Army during the 1920s was very different from what it is today and the following account is based on the experiences of Major Ernest Chapman. He was posted to Tidworth from the Cavalry Depot in Canterbury in 1927 when he was aged seventeen and joined the 3rd/6th Dragoon Guards in Assaye Barracks. A year later the Regiment was redesignated the 3rd Carabiniers (Prince of Wales's Dragoon Guards) while in 1971 it amalgamated with The Royal Scots Greys (2nd Dragoons) to form The Royal Scots Dragoon Guards (Carabiniers and Greys). As might be expected, his activities were centred almost entirely on horses and he spent the first year learning to ride for which there were ample facilities both on the Plain and in the riding school, since every cavalry regiment had its own.[51]

Reveille at 0600 hrs. was followed by Stables when the horses were fed and mucked out and this lasted until breakfast at 0800 hrs. The morning included an hour in the riding school and another hacking on the Plain while sabre drill normally formed part of the afternoon's activities. For the first three months recruits were only allowed to ride with a snaffle and single rein from which they progressed to a curb and double bridle, finally being permitted to ride with a sabre when they had reached the required standard of proficiency. Stables were held again at 1600 hrs and most of the evening was taken up with cleaning tack and personal kit but on Saturdays they always rode bare back so as to have a 'night off' from cleaning. Since there was always church parade on Sundays there was little time for relaxation other than a visit to the Ian Hamilton Home for a cup of tea and somewhere comfortable to sit, or an evening at the cinema where the programme was changed twice a week. At this time a lot of work was in hand on improvements to the Tattoo arena and on the construction of sports grounds and many afternoons were spent on digging, moving and levelling soil. On important occasions such as a Royal Visit, the whole of the 2nd Cavalry Brigade which comprised three regiments and about 1,200 horses, would be on parade. In the 3rd/6th Dragoon Guards, 'A' Squadron rode blacks, 'B' Squadron bays, 'C' Squadron chestnuts, the trumpeters greys and the band the 'odds and sods'.[52]

On 20th July 1927, the 2nd Bn. The Somerset Light Infantry which on its return from foreign service in the previous January, had joined the 7th Infantry Brigade at Tidworth, received New Colours. These were presented by HRH The Duke of York, KG, PC, KT, GCMG, GCVO, Colonel in Chief of the Regiment who in 1936 was to ascend to the throne as HM King George VI. During the 1920s consideration was being given to the lessons of the First World War and also to the problems that lay ahead, more especially in view of the continuing development of mechanization. In order to study the implications of such matters, an Experimental Armoured Force was formed and on 12th July 1928 a demonstration was held on Windmill Hill Down between Tidworth and Ludgershall, for Members of both Houses of Parliament. The programme stated that the Force which was under the command of Brig. R.J. Collins, CMG, DSO was composed of the following,

HQ Armoured Force
9th Field Brigade, Royal Artillery
9th Light Battery, Royal Artillery
17th Field Company, Royal Engineers
Armoured Force Signal Company
2nd Bn. The Somerset Light Infantry (Prince Albert's)
3rd Bn. Royal Tank Corps (less 1 company)
12th Armoured Car Company (less 1 section)
5th Bn. Royal Tank Corps
10th (MT) Company, Royal Army Service Corps.[53]

The 9th Field Brigade included one battery of self-propelled 18pr guns, known as 'Birch Guns' which were named after Lieut.-Gen. Sir Noel Birch, KCB, KCMG and were the forerunners of the modern self-propelled gun. However, by 1930 the Experimental Force had been disbanded and a great opportunity lost, the effect of which was to become painfully evident in the Second World War. Mechanization now began to gain momentum and in 1928 the first two cavalry regiments lost their horses: the 11th Hussars (Prince Albert's Own) now The Royal Hussars (Prince of Wales's Own) and the 12th Royal Lancers (Prince of Wales's) now the 9th/12th Royal Lancers (Prince of Wales's).

By 1930 Tidworth had regained much of the stability and even tenor of its pre-war days and the following is a description of the Garrison as seen by two small boys who lived in North Tidworth in the 1930s, as remembered by them. The whole of the Garrison area was immaculately kept with not a blade of grass out of place and before any very senior officer visited the barracks, the approach roads were cleaned by hosing them down. There was always something of interest going on, especially in the way of sports and youngsters would play truant from school if a particularly exciting match was to be played. Children were allowed into army gymnasiums to watch PT and the two boys were often late when running errands after stopping on the way to look at something which caught their eye. The numerous recreation grounds were always in regular use while in summer when the River Bourne dried up, it was possible for little people to crawl along the river bed into the Oval to watch anything of interest that was going on.[54]

Some of the first houses to be provided exclusively for civilian employees were built in 1927 in Nun's Walk which is now part of Sidbury Circular Road. About five years later additional houses for employees were erected in Nepaul Road adjoining Nun's Walk and all these houses were still in use in 1985. A very clear impression of the layout of Tidworth in the 1930s can be obtained from a study of the large-scale Ordnance Survey map that was published in 1936.[55] This shows each individual building, the line of the military railway as well as the names of the roads, barracks and other features.

As the international situation became more unsettled, ominous signs began to appear. The Government's decision to introduce compulsory military service came into effect with the formation of the Militia in May 1939 and this in turn led to the erection of Mathew Barracks in order to provide accommodation for the new intake. The Barracks which were presumably named after Maj.-Gen. Sir Charles Mathew, KCMG, CB, DSO, were built on land lying between the railway line and the road to Perham Down which is now occupied by the Mathew Housing Estate. Another significant change which was announced in May, was the formation of the Royal Armoured Corps[56] and as Regular Army reservists began to arrive at Tidworth during August, the peacetime atmosphere in the Garrison changed very quickly. After the declaration of war on 3rd September, the reservists were followed by volunteers who had been accepted for service and one of these was

David Niven who had left a film career in Hollywood to re-join the Army from which he had resigned in 1932. On his return, his first impression of Tidworth was not very cheering.

> It was a cold, wet, windy and altogether miserable winter's evening when I stopped the car on a hilltop on Salisbury Plain. Wood pigeons were homing into the dark isolated clumps of firs, the sodden turf on either side of the road had been chewed up by tanks. Below me lay Tidworth ... acres and acres of Victorian barrack squares in all their red-brick horror. Grey slate roofs glistened in the drizzle. A band of blue smoke hung above the place and the sound of bugles rose.[57]

Unlike the First World War, no Commonwealth units were stationed in Tidworth: the Canadians[58] and New Zealanders [59] were at or near Aldershot although they used the Plain for training, the Newfoundlanders were at Larkhill and the Australians mainly at Lopcombe Corner. While further information relating to the Newfoundlanders will be found in Chapter 10, Lopcombe is sufficiently close to Tidworth to be included in this chapter. On 17th June 1940 the convoy carrying the Australian and New Zealand contingents reached Gourock on the Clyde and after disembarking the Australian Force proceeded to Salisbury Plain, Force Headquarters being established at Amesbury Abbey and the 18th Brigade at Lopcombe Corner Camp.[60] This was located near the junction of the Salisbury–Winchester road (A30) and the Salisbury–Andover road (A343) which is known as Lopcombe Corner and lies eight miles due south of Tidworth. In addition to the 18th Brigade the troops included artillery, technical units and reinforcements and steps were taken to form another brigade from these which was designated the 25th Brigade. Although at first referred to as 'an improvised formation', it later became a permanent part of the Australian Imperial Forces.[61]

During the summer of 1940, all efforts were concentrated on dealing with a possible invasion and 18th Brigade became Southern Command Striking Force. At the same time an infantry leaders school was set up while sixty-eight officers and NCOs attended courses at the School of Artillery, Larkhill.[62] On 4th July HM King George VI inspected Lopcombe Camp and watched individual units on the parade ground and at training[63] while on 4th September, the Prime Minister visited the camp.[64] Eventually, on 16th October 1940, the Australian Force moved from Lopcombe to Colchester and a month later the 18th Brigade embarked at Glasgow for the Middle East. They were followed by the 25th Brigade on 3rd January 1941, leaving in England an administrative section and two forestry companies.[65]

In June 1942 it was disclosed that units of the United States Army would be arriving in this country in the near future and although, for security reasons no details were given, those concerned were fully aware of the considerable administrative problems that would arise from such a decision. While extensive accommodation would be needed for personnel, as well as an immense amount of storage space for the almost unlimited quantities of equipment and supplies, probably the greatest need was for a suitable training area of sufficient size. To meet such requirements one area must have come to mind immediately – Salisbury Plain – and it was here that the United States Army established itself. Consequently Tidworth Garrison, Perham Down and Warminster Barracks together with eighteen camps on, or closely adjacent to the Plain were allocated to the Americans. Of these camps, twelve lay in the vicinity of Warminster, four were in the Tidworth area and two were not far from Larkhill. The camps near Tidworth were Park House, Tidworth Pennings, Windmill Hill and Lopcombe Corner.[66] During the next few weeks the Headquarters of the United States II Corps were established at Tidworth under the command of Gen. Mark Clark.[67]

On 22nd July virtually the whole of the Garrison was taken over including the married quarters – the 'Merthyr Tydfils' – as well as the school in Bazaar Road. On 8th August 1942 the 1st Infantry Division arrived at Tidworth and was followed on 18th October by the 29th Infantry Division, while the 2nd Armored Division moved in on 27th November 1943. The 5th Infantry Division reached Tidworth on 26th January 1944 and by 1st December it had been followed by the 7th, 9th, 12th and 8th Armored Division in that order. The amount of time that each Division spent at Tidworth varied: the 29th Infantry Division remained for seven months, the 2nd Armored for six months and the 1st Infantry Division for nearly three months. Other Divisions stayed for periods of between two months and two weeks while those that arrived after D-Day (6th June 1944) generally stayed for the shortest time.[68]

In September 1942 arrangements were made for Tedworth House to be taken over by the American Red Cross as a club for American soldiers. It opened early in October under the management of Mrs. Theodore Roosevelt, Junior, whose husband, Brig. 'Teddy' Roosevelt was in the 1st Infantry Division.[69] The Club proved to be a great success and in due course special equipment was installed in the kitchens. This included a grill for making hamburgers and a doughnut machine which could produce 'several hundred doughnuts a day'.[70] This must have been very popular with the soldiers but it was seen in a rather different light by the ladies who worked at Tedworth House.

> We used to go there to make mountains of sandwiches for the troops and their cooks made more mountains of doughnuts. It was rather soul destroying when we had two ounces of margarine and one ounce of butter and one egg a week. Well, we survived...[71]

It soon became apparent that the American approach to some subjects was very different to that of the British and this was especially evident when driving a vehicle.

> The GIs seemed obsessed with speed. They would drive up to my HQ at full steam only to brake at the entrance in a cloud of dust and scream of tyres, and before the engine's fly wheel had ceased revolving the dispatch rider would be half way up the steps of the building, to deliver what would, as likely as not, be of minor consequences and seldom urgent.
>
> If a vehicle pulled up carelessly or blocked oncoming traffic the GI behind would not hesitate to overtake, even if it meant bumping up on a grass bank or knocking down anything in the way. The Americans made a shambles of Tidworth in their first winter: with the garrison roads a sea of slush and the one-time neat grass edges rutted by the wheels of the enthusiastic GIs. In contrast, Bulford, the next army establishment, run by the British was immaculate.[72]

On 3rd January 1945 the last American Division, the 8th Armored, left for France and on 19th April 1946 the remaining American units vacated Tidworth which once again became a British Garrison. On 1st December 1982 a plaque commemorating the service of the 1st Infantry Division in Tidworth, was unveiled besides the entrance to the Garrison Headquarters in Lowa Road. This was presented, on behalf of the Veterans of the 1st Infantry Division by Col. Hank Hangsterfer who had served in the 16th Infantry Regiment at Tidworth. The inscription is as follows

HEADQUARTERS
1st INFANTRY DIVISION
UNITED STATES ARMY

MAJOR GENERAL TERRY ALLEN – COMMANDING GENERAL
BRIGADIER TEDDY ROOSEVELT JR – ASSISTANT COMMANDER
BRIGADIER CLIFT ANDRUS – ARTILLERY COMMANDER

* * * *

AUGUST 1942

Although plans for modernizing Tidworth were being prepared in 1938, the unstable conditions in Europe discouraged any positive action being taken and after the War when the Army had been considerably reduced in size, many of the buildings were empty and had begun to deteriorate. However by 1949 new plans had been approved and the Ministry of Works had taken Assaye in hand, so that by the end of 1953, three barracks had been brought up to modern standards.[73] In 1984 another scheme for modernizing the Garrison was under consideration although no details had been announced by January 1985. During the post-war years a number of buildings were demolished or basically altered while others were put to an entirely different use. In 1956, a building at the junction of Bazaar Road and Jamrud Road which had previously been a canteen, was converted into a nurses home and named Cavell House. However when the Military Hospital (originally Delhi Barracks) was closed in 1977, the building was altered once again and in 1985 it became the Garrison Sergeant's Mess Annexe.

One of the most popular events in Tidworth after the Second World War, was the Tidworth Army Horse Trials which were first held in 1954. These originally took the form of a One Day Trial but in 1956 they were changed to a Three Day Event and continued as such for the next twenty-two years, being held for the last time in 1978. However, on his appointment as DAQMG at Headquarters South-West District in 1979, Maj. R.D. Baker, 16th/5th The Queen's Royal Lancers, found that the Trials were greatly missed by the local population, both civilian and military. Subsequently, with the support of the GOC, Maj.-Gen. J.H.B. Acland, CBE (afterwards Sir John Acland) and Maj.-Gen. F.E. Kitson, CBE, MC who shortly afterwards became Lieut.-Gen. Sir Frank Kitson, KCB, he re-established the trials as a One Day Event in 1980. This trial and the two that followed in 1981 and 1982 proved to be so popular, that in 1983 it was extended to cover two days, although technically a One Day Horse Trial. It is now well established and is recognized as one of the best One Day Trials held in the United Kingdom.[74]

After 1950 a considerable amount of housing development was carried out by the War Department, later the Ministry of Defence, and the Local Authority. Most of the Army houses lie in four main areas: the Zouch Estate dating from about 1961 to 1968; the Mathew Estate built between 1964 and 1968; the Wylye Road development of about 1966 and the area adjoining Plantation Road and Church Lane which was developed over a long period between about 1946 and 1967. Two housing estates were established by the Local Authority – now the Kennet District Council – the first extending from Pennings Road to Hill Top Avenue which was developed between 1951 and 1955 or thereabouts. The second estate which lies to the south of Ludgershall Road and includes Beech Hill Road, Chestnut Avenue and Hawthorn Road was built between 1965 and 1970.[75]

During its history Tidworth has been honoured by visits from many members of the Royal Family but July 1983 must be a unique month in this respect. On 5th July, HM Queen Elizabeth the Queen Mother, as Colonel in Chief, visited the 3rd Bn. The Light Infantry at Aliwal Barracks. A few days later, on 15th July, HM The Queen presented a new Guidon to the 16th/5th The Queen's Royal Lancers of which she is Colonel in Chief while on 23rd July new Colours were presented to the 1st Bn. Queen's Own Highlanders (Seaforths and Camerons) by their Colonel in Chief, HRH The Duke of Edinburgh. Finally, on the afternoon of 23rd July, HRH The Prince of Wales played for the Royal Navy in a polo match against the Army, watched by Prince Phillip.

No account of Tidworth would be complete without some reference to its trees which add so much to the appearance and amenity of the Garrison. For the most part, these are broad-leaved species comprising ash, beech, horse chestnut and lime with a small proportion of oak although recent plantings have included other species, such as horn-

(Rev. J. J. Holliman SCF)

14. Maj.-Gen. Sir Henry Rawlinson laying the foundation stone of St. Michael's Church on 22nd May 1911.

(*N. D. G. James*)

15. The Garrison Theatre which stands at the junction of Lowa Road and Bazaar Road and was opened in March 1909.

PROGRAMME of a

Concert

GIVEN BY

DAME NELLIE MELBA, G.B.E.,

to the soldiers of TIDWORTH GARRISON,

ON SATURDAY, 12TH JUNE, 1926.

1. Band ... Selection.
2. Mr. John Brownlee (*a*) "Invictus" ... *Huhn*
   (*b*) "Floral Dance" *Katie Moss*
3. DAME NELLIE MELBA
   "Jewel Song from Faust" ... *Gounod*
4. Mr. Harold Craxton, Pianoforte Solo
5. DAME NELLIE MELBA, Group of Songs.
6. Mr. John Brownlee
   Songs (*a*) "The Company Sergeant Major"
   (*b*) "The Toreador Song from Carmen" *Bizet*
7. Band ... Selection.
8. Mr. Harold Craxton, Pianoforte Solo
9. DAME NELLIE MELBA "Good-bye" ... *Tosti*

GOD SAVE THE KING.

Salisbury Times Co., Ltd., Printers, Dews Road, Salisbury.

16. Programme of a concert held at the Garrison Theatre on 12th June 1926.

(*K. Pickernell*)

(*N. D. G. James*)

17. Tedworth House: the south front.

(*J. Moloney*)

18. The Ian Hamilton Wesleyan Soldiers' Home which opened on 4th July 1908. On 18th October 1964 part of it became the Garrison Church of St. Andrew and St. Mark but this was closed in September 1985.

(*4th/7th Royal Dragoon Guards*)

19. 4th/7th Royal Irish Dragoon Guards (now 4th/7th Royal Dragoon Guards) and the Regimental Band on Church Parade at Assaye Barracks in 1912.

(*4th/7th Royal Dragoon Guards*)

20. 4th/7th Royal Irish Dragoon Guards returning to Assaye Barracks after Church Parade. 1912.

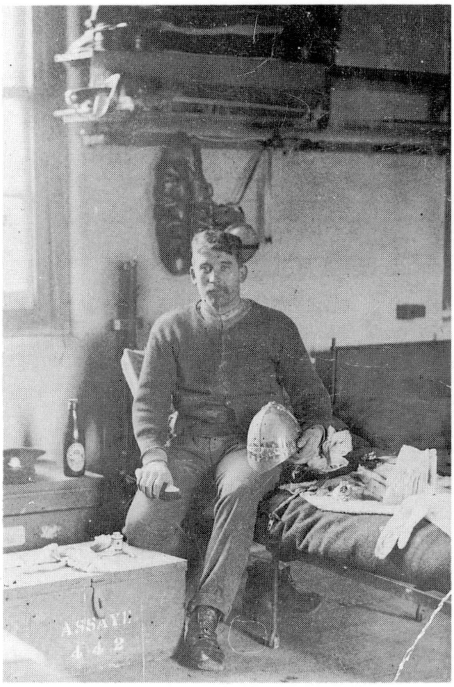

(4th/7th Royal Dragoon Guards)

21. A trooper of 4th Royal Irish Dragoon Guards cleaning his kit in Assaye Barracks in 1912. On the shelf above his head lies his sword while hanging on the wall is his bandolier, mess tin and water bottle. In a clip attached to the wall is a SMLE .303 rifle and in the foreground a barrack room box bearing the name ASSAYE and the number 442.

(L. Campbell)

22. Bhurtpore Barracks, Tidworth, 1908.

(Lt.-Col. A. Hunt)

23. Presentation of Colours to 2nd Battalion The Somerset Light Infantry by HRH The Duke of York (afterwards HM King George VI) at Tidworth 20th July 1927. Lt. A. Hunt is receiving the King's Colour.

(16th/5th The Queen's Royal Lancers)

24. The polo team of 16th/5th Lancers (now 16th/5th The Queen's Royal Lancers) before their match with 7th Queen's Own Hussars (now The Queen's Own Hussars) at Tidworth on 8th July 1926. The team comprised HM King Alfonso of Spain (second from left), Lt.-Col. G. F. H. Brooke, Maj. T. L. Horn and Capt. A. W. M. S. Pilkington.

(*Imperial War Museum*)

25. HM King George VI inspecting 2/9th Infantry Battalion Australian Imperial Force at Lopcombe Corner 4th July 1940. With His Majesty is the Commanding Officer Lt.-Col. J. E. G. Martin while behind the King is Maj.-Gen. L. Morshead, Commander, 18 Infantry Brigade.

(*Imperial War Museum*)

26. Mr. Winston Churchill with Gen. Eisenhower during a visit to American troops at Tidworth in April 1944. The officer next to Gen. Eisenhower is Maj.-Gen. E. H. Brooks, Commander, 2nd Armored Division.

(*HQ United Kingdom Land Forces*)

27. HM The Queen speaking to Capt. R. House, REME when she presented a new Guidon to 16th/5th The Queen's Royal Lancers at Tidworth on 15th July 1983. Accompanying Her Majesty is the Commanding Officer, Lt.-Col. J. A. Wright.

beam and birch. While some of the oldest trees may well have been planted by Thomas Assheton Smith who died in 1858 or his successors, it should be borne in mind that by 1985 Tidworth had been Government property for eighty-eight years and that under favourable conditions a tree can grow to an appreciable size in that time. Some of the more notable trees in the Garrison are the limes which form the very fine avenue leading from Tedworth House to Park Road; the group standing between Humber Lane and the River Bourne, opposite Jellalabad House, and the row of twenty-three horse chestnuts on the south side of Ludgershall Road between Ebble Close and Kennet Road. A large number of trees have also been planted in the area of Bulford Road, The Mall and Grand Trunk Road.

In recent years, a great deal of work has been done by the Forestry Staff of the Property Services Agency under the direction of B.A. Painter, MBE, the Chief Forester, in re-establishing and improving many of the woodland areas around Tidworth. These have included plantations and small coverts on Clavendon Hill, Chalkpit Hill and Sidbury Hill and also Andover Plantation opposite Ashdown Terrace in Tidworth. A large number of trees have also been planted in Tidworth Park to replace those that have died or have blown down.

CHAPTER SEVEN

Perham Down and Ludgershall

Perham Down lies about a mile to the east of Tidworth and a similar distance south of Ludgershall but in spite of its proximity to these two centres, it has always retained its identity. The barracks which are divided into two more or less equal parts by the Tidworth–Andover road, are of interest for two reasons. Over a period of fifty years Perham was the scene of frequent demolitions followed by successive re-building schemes which are described later in this chapter while for some seventeen years it was occupied either as a whole or in part by the same unit – the 5th Bn. The Royal Tank Corps.

The area now known as Perham Down originally formed part of Manor Farm, North Tidworth and Tidworth Down Farm[1] belonging to the Tedworth Estate which was sold by Sir John William Kelk to the War Department in 1897. From the start there was an urgent need for suitable sites on which tented camps could be erected and the matter was considered by the Salisbury Plain Committee at a meeting held on 20th April 1898.

> Proposals as to camping grounds from GOC Southern District were submitted to the Committee ... The IGF (Inspector-General of Fortifications) said that he presumed the camps referred to were the permanent camps ... The camps proposed for this year were in the neighbourhood of Ludgershall and Bulford.[2]

It is not clear from this minute whether the camp 'in the neighbourhood of Ludgershall' referred to Perham Down or Windmill Hill, as both these camps are shown on an Ordnance Survey map published in 1903 which had been revised up to 1899.[3] However, the fact that Perham Down Camp is marked on this map shows that it had been selected as an official camping ground by 1899.

With the establishment of Salisbury Plain as the principal training ground of the British Army, large numbers of troops occupied the various camps shown in Appendix IV, during the summer months. In 1906 a Field Post Office was established at Perham to deal with the mail which arose from the units which were in camp there. Over a period of seven days from 21st to 28th July, 3,500 men posted 16,227 letters which was the equivalent of 2,318 letters per day or an average of 4.6 letters per man. From 4th to 13th August, no less than 44,231 letters were written by 8,000 men, equivalent to 4,423 per day or 5.5 letters per man.[4] Whatever criticisms there may have been of the Army in those days, it could hardly be described as illiterate.

On 22nd May 1913, a review of some 8,000 troops was held at Perham Down by Gen. Sir Horace Smith-Dorrien, GCB, GCMG, DSO and the fact that the Royal Flying Corps took part made it an event of more than usual interest. As he noted in his diary

> Ceremonial parade at Perham Down. First occasion on which the Flying Corps appeared in the British Army. They flew by very well.[5]

In accordance with the policy of replacing tented camps by huts, a hutted camp was built at Perham in 1915.[6] In June 1916 the training battalions of the 3rd Australian Division arrived in the United Kingdom, details of which have been given in the last chapter, and No. 1 Training Group was stationed at Perham. Subsequently four Command Depots were established to deal with men returning from hospital and of these No. 1 Command Depot was located at Perham Down.[7]

The first major development after the War, took place in 1921 when the Anti-Aircraft Artillery School was transferred from Shoeburyness to Perham.[8] In July 1922 its designation was changed to the School of Anti-Aircraft Defence (Perham Down)[9] but in the following November it moved to Biggin Hill.[10] On 26th April 1922 the 5th Bn. The Royal Tank Corps migrated from Bovington to Perham Down where it was to remain until 1939.

> Perham Down. At last we have reached our new station ... Our camp is situated on what Shakespeare would call a 'blasted heath', but nonetheless the quarters, the air, the surrounding district and more especially the tank hangars, are all that could be desired. In short, when the keenness for fire-alarms dies down, they stop chopping and changing the furniture about, the coal issue will burn, the married officers are settled, the tank maintenance is finished and the card shooting season closes, it will be a delightful place to go on leave from.[12]

There was an overlap of about seven months between the arrival of the 5th Battalion and the departure of the School of Anti-Aircraft Defence.

> One thing does strike us as unique. Our neighbours, the Anti-Aircraft Defence School, still preserves a most warlike atmosphere with aeroplanes, searchlights, etc., functioning almost nightly.[13]

Although the hutted camp at Perham had been built in 1915, details of the accommodation were not published until 1921 when *The Barrack Book* showed that it could provide for 160 officers and 4,878 NCOs and men.[14] The Ordnance Survey map of 1926 which was revised in 1923 shows that there were three distinct areas occupied by huts and other buildings.[15] The main barrack area was divided into two halves by the Tidworth–Andover road (as it is today) each part having a frontage to the road of approximately 500yds (460m.) and extending for a depth of some 650yds (230m.) north and south of the road. The area on the north side of the road included a church and two institutes or canteens. About 500yds beyond the northern boundary of the main barrack area and running from south-west to north-east was another large hutted camp which covered an area of approximately 3.75acres (1.5hectares), the north-eastern boundary terminating at what is now Somme Road. Some 250yds to the north was a smaller camp of 1.75acres (0.75hectares) which, sixteen years later, was to form the eastern section of Fowler Barracks.

The post-war development of Perham Down began in 1922 when construction started on buildings to accommodate a tank battalion, at a cost of £32,500 while a further sum was allocated for this work during 1924–1925. In January 1926 it was noted that

> New brick married quarters are gradually growing and most of the barrack rooms have been converted.[17]

While it is not clear for what purpose the barrack rooms had been converted, the new married quarters were undoubtedly very welcome as Perham Down was already known as 'Perishing Down'. During 1926–1927 three married quarters were built for officers, one of which was Cambrai House,[18] and these were followed by another seven officers houses and additional tank hangars in 1928–1929.[19] In spite of the new buildings which were built during the 1920s, the barracks had become very much smaller as the following figures show. In 1921 accommodation was available for 160 officers and 4,878 NCOs and men but by 1929 this had been reduced to 26 and 492 respectively, owing to the demolition of the huts in the two areas to the north of the main block and to which reference has already been made.

After reviewing the 3rd Division at Dunch Hill near Aliwal Barracks on 16th May 1924, Lieut.-Gen. Sir Walter Congreve, VC, KCB was present at a 'battle' which was staged at Perham Down on the following day. The General who was about to take up the appointment of Governor and Commander-in-Chief of Malta watched the demonstration from Warren Hill. The Army units that took part were the 1st Brigade Pack Artillery, Royal Garrison Artillery, a battalion of the Wiltshire Regiment, a battalion of the King's Shropshire Light Infantry and the 5th Battalion the Royal Tank Corps. Considerable interest was added to the display by two aircraft of the Royal Air Force which 'bombed' the gun position and directed the 'fire' of the tanks on to the guns.[22]

On 1st April 1931 the 1st Tank Brigade was formed on an experimental basis and continued in operation until 30th September of that year. It consisted of the 2nd, 3rd and 5th Battalions of the Royal Tank Corps which were stationed at Farnborough, Lydd and Perham Down respectively. On 5th August, the 3rd Bn. moved to Hedge End Camp at the junction of Somme Road and the Tidworth–Ludgershall road and on 5th September the whole Brigade moved to Tilshead where training continued until 17th September.[23] In the following year another experimental brigade was formed which also carried out a period of training at Tilshead.[24] These trials contributed very largely to what was described as 'the most important development in the Royal Tank Corps since the war' when the 1st Tank Brigade Headquarters were established at Farnborough on 1st April 1934. The Brigade comprised the 1st (Light) Tank Bn. and the 5th Tank Bn. at Perham Down, the 2nd Tank Bn. at Farnborough and the 3rd Tank Bn. at Lydd.[25] During the autumn of 1934 a certain amount of re-organization took place at Perham, the 5th Bn. moving into brick built barracks and the 1st (Light) Bn. moving into the huts which the 5th had vacated. At the same time approval was given for the whole brigade, except the 2nd Bn., to move to Perham so as to form 'a self-contained "Tank Town"!'[26]

The following recollections of life at Perham Down between the wars have been kindly provided by Mrs. M.B. Berkery and Mr. R.V. Peters, whose fathers served in the 5th Bn. Royal Tank Corps at Perham Down during that period. In 1928 most of the buildings were constructed of timber and corrugated iron and the married quarters usually consisted of a hut which contained a kitchen, living room and three bedrooms with a passage on one side to give access to each room. Cooking was done on an iron range in the kitchen while the living room and the bedrooms were heated by small stoves known as 'Queenie Stoves', all of which were coal fired. Bread, meat and coal were drawn in bulk by the Battalion Quartermaster and then issued to families in accordance with their size. The children attended school at Perham and two events which they all looked forward to were the school sports day, held during the summer, and the children's Christmas party which was organized each year by the Battalion and which was always a very happy occasion.

Church parades were usually held at the Garrison Church in Tidworth and those attending would march there and back. When the Battalion moved to Tilshead for the summer camp which lasted for about six weeks, the families remained at Perham. If, however, they were out of sight they were certainly not out of mind as during each camp the Battalion held a Families Day to which everyone, especially the younger members, greatly looked forward. After arriving at Tilshead in coaches which were provided by the Battalion, there would be a programme of sports, games and other entertainments with a special dinner at about midday and tea later in the afternoon. Altogether, it was a day to remember. Another occasion which is unlikely to be forgotten by those who were present, was a magnificent fireworks display given by the contractors, Messrs W.E. Chivers & Sons in 1936, to mark the completion of the rebuilding of the barracks at Perham. However the most important social event of the year was the Cambrai Ball which was held on the Friday or Saturday nearest to Cambrai Day – 20th November – and was attended by all ranks.[27]

Mr. L.E. Trevett who, from 1931 to 1938, also served in the 5th Battalion has kindly supplied the following information. Each year at the end of March or the beginning of April, the Battalion moved to Lulworth for gunnery practice where it lived under canvas. At the end of April it returned to Perham where it stayed until June when the whole Brigade moved to a tented camp at Tilshead where it remained for about six weeks.[28] Perham Down had its own polo ground which lay to the south-west of Newdrove Plantation and until the beginning of the Second World War, polo was played on it every Saturday afternoon during the summer. After the War less use was made of the ground although the occasional game was still played until 1984. It is now in use as a general sports field.

By 1932 work was in hand on another building programme and photographs in the Tank Museum Library at Bovington show 'the new barracks under construction' in August 1933, 'nearing completion' in November and almost finished in February 1934. Approval was given for the provision of accommodation for a second tank battalion during the year 1934–1935[29] and in April 1935 it was noted that

> The rebuilding of Perham is going on apace and the latest addition to its many wants which will be greatly appreciated by everyone, is the laying down of a concrete tank park.[30]

Two officers' messes were built in 1935[31] one presumably for the 1st (Light) Battalion which had moved to Perham in 1934 and the other for the 5th Battalion. In 1985, the building which is nearer to Tidworth was an officers' mess while the further one was a warrant officers' and sergeants' mess although at some time previously, the position had been reversed. In spite of the fact that the question of moving 1st Tank Brigade Headquarters to Perham, had been under consideration for some time, it was not until May 1937 that this took place.

> The temporary accommodation now being reconstructed for Tank Brigade HQ is nearing completion, so we hope to welcome them to our midst before this appears in print.[32]

Early in 1938 it was noted that 'the sudden orders for the move of 1st (Light) Battalion to Egypt came as a great surprise to us all'[33] and on 18th March the Battalion entrained at Ludgershall for Southampton.[34] However it was not long before the resulting vacancy at Perham was filled.

> 16th May 1938 saw the rebirth of the 8th Bn. ... now busy forming at Cachy Barracks, Perham Down, as an Army Tank Battalion and comes under the orders of 3rd Division ... We are very fortunate in taking over excellently equipped barracks constructed on modern lines. Let us hope that we shall remain for a considerable time in them.[35]

In fact the 8th Battalion remained at Perham until a few days before the German army launched their attack in mid-May 1940.[36]

As the various building projects were completed, Perham Down reached a state of comparative stability and the following description of the barrack area as it was in 1936, is based on the Ordnance Survey map which was published in that year. The connection between Perham and the Royal Tank Corps was emphasized by naming the barracks, roads and senior officers quarters in commemoration of battles and engagements of the First World War in which tanks had taken part. The main barrack area was divided from north to south into two parts with Cambrai Barracks to the east and Cachy Barracks to the west. Little need be said about Cambrai except that it was near this town, on 20th November 1917, that the Royal Tank Corps achieved its greatest triumph in the First World War. Near Cachy, a village some three miles south of Villers-Bretonneaux, seven whippet tanks of the 3rd Battalion overran and decimated two German battalions on 24th April 1918.[38] The names of the roads included those of Bermicourt, Cachy, Drocourt, Fremincourt, Flers, Fontaine and Kemmel while Somme Road connected Perham with Ludgershall as it does today and by which name it was still known in 1985. It was at Flers, on 15th September 1916, that tanks were used for the first time.

Each barracks had its own officers' mess, warrant officers and sergeants' mess, guard room and institutes and while a church was shown on the Ordnance map of 1926 near the junction of Somme Road and the Tidworth road,[39] it does not appear on the 1936 map. On the south side of the main road, opposite Somme Road, were the soldiers married quarters which included the Infant and Senior Schools. To the west of Cachy Barracks in Lambdown Road (now known as Lambdown Terrace) were the officers' quarters while beyond these were the senior officers' houses – Cambrai House, Cachy House, Hamel House and Bourlon House – which were still so named in 1985.

During 1938 two new camps were built on either side of Somme road, about 600yds (550m.) to the north of the main barracks and adjoining what is now, the south-western boundary of the RAOC Vehicle Depot. The camp on the west side of the road was named Fowler Barracks, presumably after Lieut.-Gen. Sir John Fowler, KCB, KCMG, DSO who was Colonel Commandant of the Royal Corps of Signals from 1923 to 1934. On the east side, and opposite Fowler Barracks were Busigny Barracks named after a village in France which is situated between Le Cateau and St. Quentin. Fowler Barracks were demolished during the mid 1960s and Busigny which closed in 1976, were pulled down shortly afterwards although the camp cinema remained until 1984.[40]

Although the 5th and 8th Battalions were still at Perham in April 1939[41] the Second World War brought many changes and by May 1940 both battalions had gone. In 1940, the 54th Training Regiment RAC was formed at Perham but in February 1944, after it had moved to Catterick, the 8th Cavalry Squadron and three artillery battalions of the 5th Armored Division, United States Army, arrived at Perham Down.[42] They remained there until June when they left in order to take part in the invasion of Europe. At the end of the War, Perham became a transit camp for those who had married American soldiers – generally referred to as 'G.I. Brides' – and who were awaiting a passage to America. By February 1946, this operation was in full swing.[43] During the post-war years numerous regiments have been stationed at Perham Down including the 3rd Regiment RHA, the 11th Hussars, the 13th/18th Royal Hussars (QMO), the 5th Field Regiment RA, the 32nd Assault Regiment RE and the 3rd Division Airhead Ordnance Company RAOC.

The first brick-built married quarters were erected on the south side of the Tidworth–Andover road, at varying times between 1927 and 1961[44] and in 1985 the roads serving these quarters bore the names of battles of the Second World War. During 1968 additional quarters were built on the north side of the main road which are known as the Canberra Estate since the roads are named after Australian cities.[45] The most recent housing development took place in 1977 with the building of Upnor Close and Wouldham Close.[46]

The final demolition and rebuilding of the barracks began in 1972, exactly fifty years after the first reconstruction programme had started in 1922. The work was completed in 1974 and during January and February 1975 the 22nd Engineer Regiment, Royal Engineers moved into Perham which had been re-named Swinton Barracks, after Maj.-Gen. Sir Ernest Swinton, KBE, CB, DSO. This was a very happy choice as not only had he been commissioned into the Royal Engineers but he was also the originator of the tank and had raised the Heavy Section of the Machine Gun Corps in 1916, under which name the first tanks operated. He was also Colonel Commandant of the Royal Tank Corps from 1934 to 1938 and in this way the link between Perham Down and the Royal Tank Regiment which existed for so many years, has been retained.

In early times Ludgershall was principally known for its castle which is first mentioned in the reign of King Stephen (1135–1154).[47] However, by 1542 it was in ruins as John Leland the Antiquary noted on his journey through Wiltshire in the middle of the sixteenth century

> Luggershaull sumtyme a castle in Wileshire 10 miles from Marleborow, and a 4 miles from Andover almoste in the way betwixt. The castell stoode in a

> parke, now clene doune. There is of late times a pratie lodge made by the ruins of it, and longgithe to the King.[48]

Over the years the name of the town has been spelt in several ways and in addition to that used by Leland, Speed shows it on his map of 1610 as Lurgishall.[49]

As time passed Ludgershall remained a small country town serving the needs of the neighbourhood but with only limited contacts beyond its immediate surroundings. However all this was to change when on 5th February 1883 the Swindon, Marlborough and Andover Railway was opened and Ludgershall obtained a railway link which placed it in direct communication with the rest of the country.[50] In June of the following year the railway amalgamated with the Swindon and Cheltenham Extension Railway and became the Midland and South Western Junction Railway[51] and continued to operate until 1st July 1923 when it was taken over by the Great Western Railway.

The connection between Ludgershall and the Army began in the summer of 1897 when the War Office acquired Manor Farm which was also known as Castle Farm, on account of its proximity to the castle ruins. It extended to an area of some 666 acres (270 hectares) and cost £9,350, the purchase being completed on 31st October 1898.[52] During the next few months consideration was given to the purchase or exchange of several small areas of land near the town, in order to round off the boundary of the estate. When the acquisition of one of these areas was being considered by the Salisbury Plain Committee at a meeting on 2nd February 1899, Col. R.M. Barklie who was responsible for such matters

> informed the Committee of the probable use that the ground south of Ludgershall would be put to (in the way of AOD depot, ASC warehouses, railway sidings and lines for distribution) and also stated that there was no probable use for the land north of Ludgershall.[53]

In spite of this statement, no buildings were to be erected on 'the ground south of Ludgershall' for another forty years. The letters 'AOD' refer to the Army Ordnance Department (now part of The Royal Army Ordnance Corps) and 'ASC' to the Army Service Corps (now the Royal Corps of Transport).

Although a branch line to Tidworth had been opened for traffic in 1902, Ludgershall station was used for important occasions. On 8th November 1917, HM King George V travelled to Ludgershall in the Royal Train and then rode on his charger to Perham Down to carry out an inspection. Some ten years later, Members of Parliament also travelled to Ludgershall station when they attended a demonstration of the Experimental Armoured Force on 12th July 1928. It was inevitable that the coming of the railway, followed by the Army would result in major changes to Ludgershall and in 1907 A.G. Bradley observed that

> within my memory it too was a quiet village, but is now the busy railway entrepot for the military section of Salisbury Plain.[54]

Between the wars the area to the south of Ludgershall remained as agricultural land although a site adjoining the Tidworth road and, what is now Somme Road, was used as a tented camp which is shown on the Ordnance Survey plan of 1936 as Hedge-end Camp.[55] However it was not until the Second World War was imminent that any steps were taken to develop the land between Simmonds Road on the southern outskirts of the town and Somme Road. The first two buildings were erected in 1939 near Hedge-end Plantation as a Mobilization Depot and shortly afterwards No. 1 Impressed Vehicle Camp was established on the site. From 1941 to 1943 it became a Clothing Sub-Depot, Command Ordnance Depot and during these years facilities were provided for the Royal Australian Army Ordnance Corps during 1941, a Royal Army Ordnance Corps Command Workshop in 1942 and the United States Army and Navy Workshops from 1943 to 1944. It was in the course of these last two years that the three massive sheds, numbers 93, 97 and 100 were erected by the American Army to facilitate the preparation of vehicles needed for the invasion of Europe.[56] It is said that crates of unassembled 'jeeps' arrived at one end of the shed, were unpacked, assembled, checked and then driven out at the other.

To facilitate the transport of tanks, vehicles and other heavy equipment, the United States Army built a railway connection in 1943, to what was then the Ludgershall–Tidworth line and this was still in use in 1985. It leaves the original branch line at a point opposite the Defence Medical Equipment Depot and after 400yds (365m.) it bears to the south over the Ludgershall–Tidworth road (A3026) by means of a level-crossing. Some 280yds (256m.) beyond this crossing the line turns to the north-east and continues for a further 690yds (631m.), the whole layout being in the shape of the letter 'U'. The total length of the branch is about 1,670yds or 0.95miles (1,526m. or 1.53km.) but the actual length of track, after taking into account passing loops and sidings is approximately 2,470yds or 1.4miles (2,260m. or 2.26 km.).

In 1943 the Clothing Sub-Depot was replaced by a Command Ordnance Depot and after D-Day a prisoner of war camp was established at the western end of the Depot. During 1946 it became a Returned Store Depot but in the course of the next year it was taken over by a detachment of No. 1 Armoured Fighting Vehicle (AFV) Depot, to be followed by No. 82 AFV Depot in 1952, No. 13 in 1953 and a Command Vehicle Depot in 1958. During 1947 a detachment of No. 1 AFV Depot Workshops, Royal Electrical and Mechanical Engineers became a sub-unit, later to be replaced by No. 11 and No.

12 Group Workshops, REME and in 1961, by No. 96 Vehicle Depot Workshops, REME. In 1971 it became the Vehicle Depot, Royal Army Ordnance Corps but this title was changed to the Armoured Vehicle Sub-Depot Ludgershall on 10th May 1985. It is the only depot in the United Kingdom which stores 'A' vehicles, that is, armoured fighting vehicles which carry 'offensive armaments' as opposed to 'B' vehicles which are load or passenger carrying: both types may be tracked or wheeled. Its work is mainly concerned with the inspection, repair and modification of vehicles and certain types of equipment but it also includes a Technical Training Wing for training the drivers of 'A' Vehicles.[57]

In 1939 the Army Medical Stores which had previously been located at Woolwich, were transferred to Ludgershall where they occupied a site some 600yds (548m.) due north of what is now the RAOC Vehicle Depot. The title was changed in 1951 to the Army Medical Equipment Depot and, in 1969, to the Defence Medical Equipment Depot. The second alteration was due to the fact that until 1968 the Royal Navy, the Army and the Royal Air Force each had their own separate organizations for providing supplies of medical and dental stores. However in 1969, the responsibility for obtaining medical stores and equipment for the three Services passed to the Medical Supply Directorate through its agent, the Defence Medical Equipment Depot, which was no longer solely concerned with supplies for the Army.[58]

The main purpose of the Depot is to provide the Services, in whatever part of the world they may be, with medical equipment and supplies of all kinds in both peace and war. This involves the purchase of medical and dental materiél, its packaging, storing, inspection and maintenance. In 1985 the staff which included an officer from the Royal Navy and one from the Royal Air Force, was composed of both military and civilian personnel in the proportion of approximately one military employé to three civilian. The Depot also trains medical storemen of the Royal Army Medical Corps, pharmacy technicians from each of the Services and a limited number of overseas students. Over the period from 1971 to 1982, the Depot was rebuilt and now comprises an extensive area of modern buildings with its own railway connection as well as road access.[59]

At Everleigh, some four miles north-west of Ludgershall and adjoining the road to Upavon (A342), are the David Bruce Laboratories which operate under the direction of the Royal Army Medical Corps. The Laboratories which are concerned with the preparation of enteric vaccines, owe their origin to the work of Sir Almroth Wright of the Army Medical School at Netley near Southampton in 1896. This work was subsequently continued in the Vaccine Department of the Royal Army Medical College at Millbank, London.[60]

In 1939, as the situation in Europe became more threatening, it was decided to move the Vaccine Department to Tidworth and on 1st

September 1939 an Emergency Vaccine Laboratory was established there. The demand for vaccines increased tremendously after the beginning of the War and it soon became evident that larger quarters would have to be found. Early in September 1939, the War Office had requisitioned Everleigh Manor as a military hospital but when the need for more laboratory space became essential, the Manor provided a very acceptable solution. Consequently, on 1st April 1942, a Subsidiary Emergency Vaccine Laboratory was opened at Everleigh which was joined by the remainder of the Vaccine Laboratory on 25th July. In the same year a temporary hospital was built close to the Manor for the United States Army but at the end of the War, this was demolished and the site restored to agricultural land.[61]

During 1946, the name of the Laboratory was changed to the Army Vaccine Laboratory but in January 1951 it was re-named the David Bruce Laboratories in honour of Maj.-Gen. Sir David Bruce, KCB who after joining the Army in 1883, served in Egypt and Malta where he discovered the cause of Malta Fever. He then spent a year in Berlin studying bacteriology. In 1894 he was posted to South Africa where he carried out a considerable amount of research into diseases caused by the tsetse fly and later took part in the defence of Ladysmith. He died in 1931. In 1985 the Laboratories were primarily concerned with the preparation and issue of vaccines for the prevention of typhoid, paratyphoid, plague and cholera. They also provide facilities for research, act as a reference laboratory for the identification of various enteric organisms and run courses for officers and laboratory technicians.[62]

CHAPTER EIGHT

Bulford – The Curtain Rises

At the end of the last century, Bulford was a small compact village which was largely dependant on farming for its existence. On the west it was bounded by the River Avon; on the east by Milston Road and Station Road; on the north by the Old Coach Road and to the south, by the road that led to Seymour Farm, now known as Watergate Lane. There were no buildings beyond these boundaries except for the Vicarage at the junction of Milston Road and the road now named Camp Road, and a building near Crescent Road.[1] Due east of the village and a little over a mile, across open country, lay the centre of what was to become Bulford Hut Barracks.

A hundred years earlier, Beacon Hill which stands some one and three-quarter miles to the south-east of Bulford church, had been selected as one of the principal stations in the proposed national survey. By 1791, it had been decided that the Board of Ordnance should undertake a survey of the whole country with a view to producing a series of accurate maps which subsequently became known as the Ordnance Survey. The work was carried out by means of a system of triangulations which were built up on a number of selected base lines, laid out in different parts of the country and measured with great accuracy. One of these bases was established between the top of Beacon Hill and a point (grid reference 144329) on the east side of the Salisbury–Amesbury road, a little to the north-east of Old Sarum. Each end of this base which is approximately 6.9miles (11.1kilometres) in length, was marked by the barrel of an old gun, partially sunk in the ground with the muzzle pointing upwards. The two sites were marked on the Ordnance Map by the words 'Gun, End of Base' but in the case of Beacon Hill, the gun was removed many years ago and these words did not appear on the large scale maps of 1901 or on later editions.[2]

However the Ordnance map (scale 1:50,000) which was published in 1979,[3] still marks the end of the base at Old Sarum and in 1967 a commemorative stone was erected on the side of the Salisbury road bearing the following inscription.

In 1794
a line from this site
to Beacon Hill was measured by
Capt. W. Mudge of the Ordnance Survey
as a base for the triangulation of
Great Britain
Presented by Master Masons 1967.

When the first Ordnance Survey maps of Salisbury Plain were issued, they were inscribed 'Published 14th August 1817 by Col. Mudge' who it may be presumed was the same officer whose name is recorded on the stone.

The first land that the War Office acquired at Bulford was an area of 751 acres (304 hectares) which formed part of an estate belonging to Miss Seymour. This purchase was completed in March 1898 but in the summer of 1900 a further 288 acres (116 hectares) were bought from her. Another property that was acquired in March 1898, was the Bulford Manor Estate covering some 1,917 acres (775 hectares) which belonged to F.C. Ledger Hill and this was followed by the purchase of an additional 2,205 acres (892 hectares) situated in and around Brigmerston, from S.S. Rendell in November 1899.[4]

Although the purchases of the Bulford Manor Estate and the land belonging to Miss Seymour were not completed until the end of March 1898, the War Office was apparently granted earlier possession, according to the following report which appeared in the *Salisbury and Winchester Journal* on 10th July 1897.

> The War Office have lost no time in taking advantage of the acquirement by the Government of an extensive tract of land on Salisbury Plain, for the purposes of a military training ground and the past week has witnessed the encampment of the 4th Cavalry Brigade at Bulford. The camp ... is situated on Bulford Down about a mile and a half to the north-east of the village of Bulford and about twelve miles from Salisbury. Beacon Hill rises close to the eastern side of the camp ... and on the western side is Nine Mile Stream.[5]

In 1985 it is difficult to realize the effect that the appearance of a large number of troops must have had on this very quiet rural area where normally a single stranger would have been a subject for comment.

> It is long since the old road from Salisbury to Marlborough ... was the scene of so much traffic since the cavalry camp was formed at Bulford for the brigade consisting of the Scots Greys and the 1st (King's), 3rd and 7th Dragoon Guards.[6]

The site which had been chosen for this first camp seems to have met with general approval.

Officers and men have both expressed themselves as much pleased with the Bulford camping ground and it is certainly about as healthy, pleasant and convenient a place as a body of troops could desire for pitching their tents.[7]

Fortunately some accounts of life at Bulford when it was still only a tented camp, have been provided by Mary Hamilton whose husband, a captain in the Royal Irish Fusiliers, was stationed there in 1900. They lived in rooms at Milston Farm and after hiring a dog-cart from the landlord of the 'Rose and Crown' in Bulford village she was able to explore the surrounding countryside.

I watched the beginnings of the camp at Bulford; the first bell-tents going up; then expanding by degrees into the vast panoramic spread of white against the green backcloth of the plain. Very different from the Bulford of 1946 with its brick and concrete and corrugated iron. Most of the troops at Bulford belonged to Militia, Volunteer or Yeomanry regiments. The infantry were nearly all in scarlet and the Yeomanry in blue, but khaki was being issued to the drafts in training for South Africa, and some of the officers and men during that hot summer were wearing the wide-brimmed Australian hats turned up at one side.

Occasionally I would hear firing on the rifle ranges, or bugles in the camp, but there was no traffic in the air and not a great deal on the roads. I would drive for miles and meet only a few farm carts, or perhaps an Army Service Corps waggon.... On Sunday mornings I would drive to the camp for church parade and attend divine service from the dog-cart. I would hear the fifes and drums and the brass bands of the different regiments assembling for the parade. The officers and men would kneel on one knee and come up to their feet for the hymns, and for the short sermon preached by the chaplain with his surplice fluttering in the breeze.[8]

An Ordnance map published in 1903 which marked the land belonging to the War Department, shows that the main camp was divided into two parts – Bulford Hut Barracks and Bulford Camp.[9] The boundaries of these two areas are shown on the map on page 94. The site on which the hut barracks were built was divided by three roads which followed the present course of Haig Road, Wing Road (previously Grierson Road) and Rawlinson Road. The land on each side of these roads was divided into eight 'lines', seven of which were identified by the letters A to G, the remaining one being known as the ASC lines since they were occupied by the Army Service Corps which in November 1918 became the Royal Army Service Corps. Further information regarding the exact location of these lines and the names which were subsequently allocated to them, will be found in Chapter 9 and summarized in Appendix VII.

The construction of the barracks which were officially designated as 'hutting', began in 1900[10] and a substantial amount of the work had been completed by November 1901.[11] Although at first the buildings were described as 'wooden hutments', it would appear that subsequently the method of construction was altered, and huts with roofs

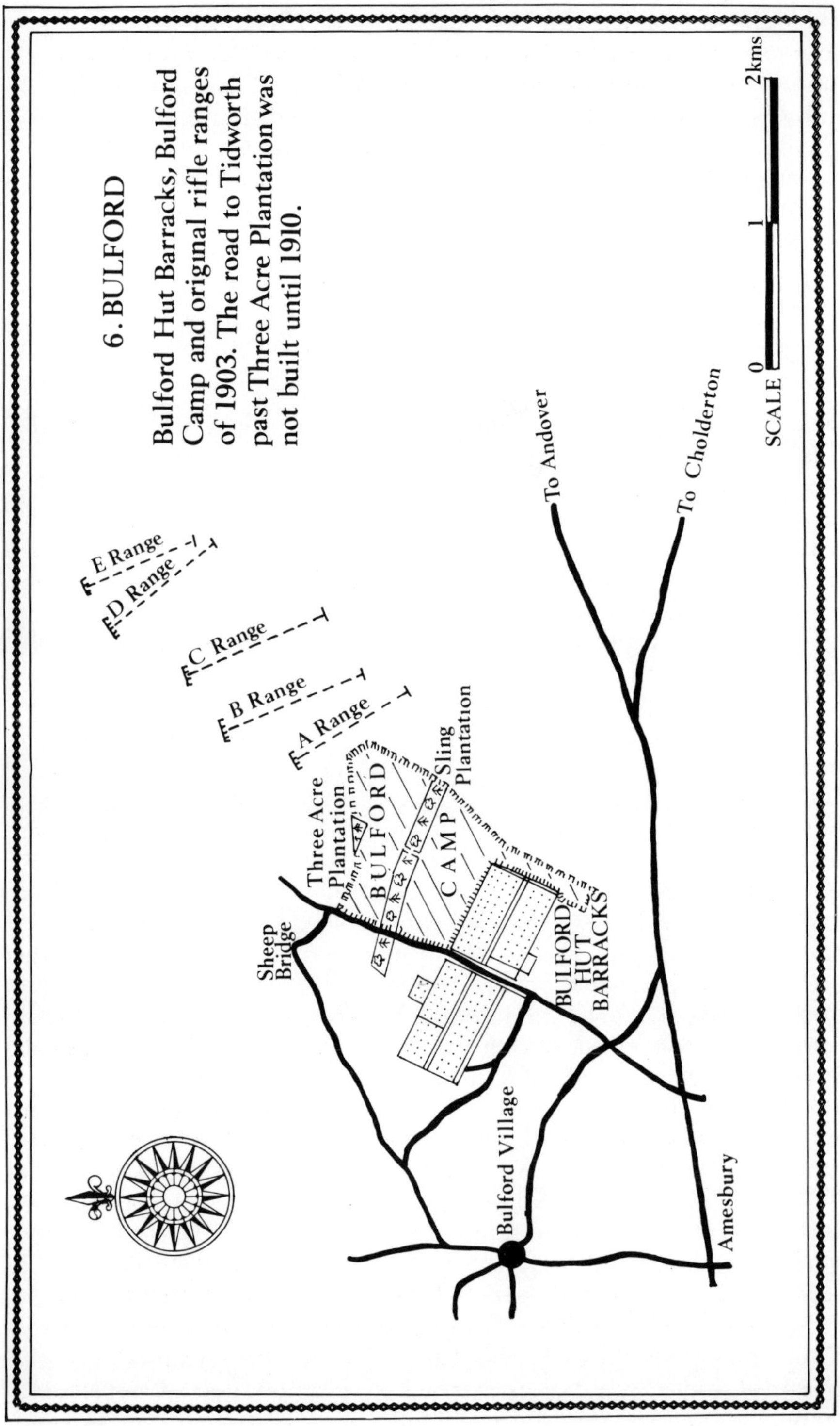

6. BULFORD

Bulford Hut Barracks, Bulford Camp and original rifle ranges of 1903. The road to Tidworth past Three Acre Plantation was not built until 1910.

and walls of wooden framing covered with felt and corrugated iron were erected. The completion was also recorded on 10th November 1901, of five warrant officers' quarters consisting of two-storeyed houses built of brick with slated roofs.[12] It has been said that some of the huts erected at Bulford were originally intended for use in South Africa during the Boer War and this may have been the case.[13] Mary Hamilton has mentioned that drafts for South Africa were in training at Bulford and she noted that after October the weather

> became cold and damp and conditions were severe in the camp. The canvas bell-tents, pleasant enough in summer, were inadequate protection against hard weather. Early in November some men died of pneumonia and it was decided to break up camp.[14]

In such circumstances it would seem reasonable to divert some of the huts in order to provide better accommodation for the troops who were to be posted to South Africa and to enable their training to be continued during the winter, if this was thought necessary.

The development of the camp continued and the local press remarked that by the autumn of 1903 'the number of timber and tin habitations have grown during the year'.[15] Some idea of the impact that the Army must have had on the neighbourhood in the early days can be gathered from the fact that in 1901 the population of Bulford was 1,386 of whom no less than 1,043 were Army personnel.[16] During the period 1904–1905, the contractors, Kirk and Randall, were paid £60 compensation in respect of the construction of 'married quarters, riding schools, etc.'[17] The three brick-built riding schools were still standing in 1985 although by that time, they had been put to other uses. Two were situated on the east and west side of Mons Avenue at its junction with Haig Road, that on the east being occupied by the Bulford Study Centre and on the west by the NAAFI Messing Store. The third, at the corner of Rawlinson Road and Provost Road was used as a gymnasium by the United Kingdom Land Forces, School of Physical Training. During the Second World War the building which is now used by the Study Centre was converted into a concert hall for the Entertainments National Services Association more generally known as 'ENSA'.

The following information which relates to the layout and organization of Bulford Camp between 1902 and 1904, has been obtained from a rather unusual source – the work books kept by the Garrison Plumber. These hand written reports are day-to-day records of the work undertaken by the plumber in different parts of the Garrison and precise details are given of each task as well as the place in which it was carried out.[18] A, B, C and D Lines were all occupied by brigades of the Royal Field Artillery, E Lines by the Royal Engineers and the 'ASC' Lines by the Army Service Corps. Each of these six lines were self-contained and in addition to barrack accommodation, they included regimental offices and guardrooms, officers' and sergeants' messes and married quarters for officers and soldiers. A and B Lines each included a school within their boundaries, that in A Lines being situated at the junction of Bulford Road and Bond Street while in B Lines the school stood between Bond Street and Milne Road and faced Haig Road. Laundries are only mentioned in connection with A, B and C Lines and if these were the only ones provided, personnel from the other lines were presumably allowed to use them. In October 1902 repairs were carried out in the office of the CRA which was in B Lines but as no reference is made to it after December of that year, it had probably moved to another location.

Occasionally the work book contained references to the Infirmary Stables and these are likely to have been located in F Lines as the Ordnance map of 1939[19] shows Veterinary Street (the southern continuation of Signal Street) forming the eastern boundary of what had been F Lines. It would also appear that in 1904 part of these lines were taken over by the Sappers in addition to their occupation of E Lines. As well as being responsible for repairs to the various water systems and their fittings which existed in the Garrison area, the plumber also had to deal with any other work which came within his trade, should the need arise. An entry for 10th March 1904 records 'repairing copper pipe for steam roller' while on 13th October he had to turn his hand to an engine of a rather different kind – 'repairing beer engine in Sergeants' Mess, C Lines: new leather to pump'.

In 1984, Mr. E. Hall of Priory Cottages, Bulford, was able to recall, at the age of 93, the time that he had spent in the Royal Field Artillery at Bulford Camp, before the First World War. Most of the buildings were 'tin huts', lit by paraffin lamps while those which formed the 'barrack rooms' usually accommodated ten or twelve men and were heated by a single stove. In 1910, he left Bulford and went to India with the 25th Brigade, RFA which was equipped with 18pr. and 4.5in. howitzers. After serving in Mesopotamia where he was taken prisoner by the Turks, he returned to England in 1919 and retired from the Army in 1931, just as his Brigade was being mechanized.[20]

In South Africa the Boers had operated as mounted infantry from the start thus greatly increasing their mobility, and eventually the British Army followed their example. This form of warfare must have created a deep impression, as more than a year after the end of the Boer War, it was reported on 9th May 1903, that

> In future three schools of instruction for mounted infantry will be maintained in the United Kingdom – one at Aldershot, another at Salisbury Plain and a third in Ireland. For each school a permanent staff will be provided. A mounted infantry battalion was formed at Bulford Camp this week.[21]

Bulford was subsequently chosen as the location for the School on the Plain but after three years it closed on 1st July 1906.[22]

In order to deal with the increasing number of troops that were moving to and from Salisbury Plain, a branch railway was built from Grately on the main Basingstoke–Salisbury line of the London and South Western Railway, to Amesbury. This was opened to the public on 2nd June 1902 and it was subsequently reported that

> No time is to be lost in continuing the L. & S.W. Railway from Amesbury to Bulford Camp and in a few years time, there is sure to be a great alteration in the Bulford and Amesbury districts.[23]

The Bulford extension which was opened on 1st June 1906, is described in Chapter 13. In the meantime training continued and the following account which appeared in the *Andover Advertiser* at the end of the 1903 training season, shows that new ideas supported by what was then unconventional equipment, were now being put into practice.

> There has been a great deal of work done; artillery, cavalry and infantry have been instructed in the modern art of warfare; the ASC (Army Service Corps) have become further familiarized with traction engines, steam motors and lorries ... The Motor car and cycle have been much in evidence; while wireless telegraphy, pontoons, and other queer craft have put in an appearance.[24]

Although one of the principal reasons for purchasing land on Salisbury Plain was to acquire an area where artillery could practice,[25] the more urgent concern of the Salisbury Plain Committee was, in fact, the provision of rifle ranges. The Secretary of State for War, Lord Lansdowne, took considerable interest in this matter and at their meeting on 7th December 1897, the Committee were informed that 'Lord Lansdowne was very anxious to have some ranges laid out at Salisbury Plain'.[26] The need for these can be more readily appreciated if the developments in rifle design which took place in the latter half of the nineteenth century, are taken into account.

The demand for a breech-loading rifle was met to some extent, by fitting a breech block devised by an American, Jacob Snider, to the muzzle-loading Enfield rifle.[27] This became known as the Snider rifle although it was, more correctly, the Snider-Enfield and was issued to the British Army in 1866. It was, however, only a temporary measure and was replaced in 1871 by the Martini-Henry which was an infinitely better weapon. This rifle had a flatter trajectory, increased range and greater accuracy although it was liable to jam under desert conditions and to excessive fouling of the bore.[28] In 1888, it was replaced by the Lee-Metford rifle which had been designed by an Englishman, W.E. Metford, and incorporated a bolt action and magazine produced by a Scotsman, J.P. Lee who had become a United States citizen.[29] With a calibre of 0.303ins (7.7mm.) it was sighted for a range of 1,800yds (1,645m.) but if special sights were fitted, it could attain 2,900yds (2,650m.).[30] After the Royal Small Arms Factory at Enfield had carried out certain modifications to it in 1895, it was re-named the Lee-Enfield and was followed in 1900 by the Short Magazine Lee-Enfield which remained in service for many years.

One result of the greatly increased range of the Lee-Metford was to bring into question the safety of existing ranges and although some were no doubt suitable, others were not. On 26th January 1897, three days before the Under-Secretary of State, the Rt. Hon. W. St. John Brodrick, announced the Government's decision to acquire land on Salisbury Plain, a Member asked a question in the House regarding the Lee-Metford rifle. Robert Wallace wished to know the position regarding this rifle which was to be used at Bisley that year, since there were only a few ranges in Scotland that were suitable for its use. The Under-Secretary replied that until longer ranges could be provided, the introduction of this rifle would cause difficulties in many different localities throughout the country.[31] The position visualized by Wallace was actually rather worse than he appreciated, as the Lee-Enfield which had replaced the Lee-Metford in 1895, used a cordite propellant instead of black powder and this gave it a still longer range.[32] It is possible that this Parliamentary question was one of the factors which contributed to Lord Lansdowne's anxiety regarding the lack of rifle ranges on the Plain.

The first occasion on which the Salisbury Plain Committee considered the question of ranges was on 18th November 1897, when it was reported that a site for a range had been selected on Warren Farm. The house and buildings of this farm lay about 1,100yds (1,000m.) south of Perham Down at grid reference 253473. Later, on 7th December, the Inspector-General of Fortifications expressed the opinion that a site adjoining Sidbury Hill would be more suitable and asked how many ranges were needed when it was agreed that two ranges 'should be sufficient for the present'.[34] As difficulties arose over the question of obtaining possession of the sites, little progress had been made by 20th January 1898.

> The Chairman said that Lord Lansdowne was most anxious that some ranges should be ready for use on Salisbury Plain this year. Parliament had voted half a million for ranges, and very little had been done in that direction so far.[35]

Following a lengthy discussion, it was agreed that one range should be built during 1898. After this meeting the minutes only contain two brief references to ranges until a meeting held on 24th January 1901 when it was reported that the construction of 'Perham Down Rifle Range' had made it impossible to use the buildings at Warren Hill and Bedlam.[36] The latter adjoined the south-eastern end of Shipton Gorse which has since been renamed Bedlam Plantation.

A map which was published by the Ordnance Survey in 1903 under the title of 'War Department Land on Salisbury Plain', showed fifteen rifle ranges and carried a marginal note which read 'Rifle Ranges inserted 1900'. From this it may be assumed that the ranges marked on the map had been constructed by that year, and probably in 1898 or 1899. These were situated in four localities, Larkhill, Bulford Camp, Tidworth and Perham Down and details of these are given in Appendix VI. Of these ranges, only those at Bulford and Perham were still in use in 1985 by which year a number of changes had been made. In 1900 there were five ranges at Bulford which were situated in the same area as those which were in use in 1985, that is, adjoining the Bulford–Tidworth road between Three Acre Plantation and The Belt. The ranges were lettered A to E, from west to east, ranges A and D being 900yds (823m.); B and C 1,000yds (914m.) and E, 800yds (731m.). In every case, the firing points were sited on the south-east side of the present-day road from Bulford to Tidworth, that is to say on the right-hand side of the road when travelling towards Tidworth. Originally this did not cause any inconvenience as the road was not built until 1909–1910.[37]

By the time the Ordnance map of 1925 had been published a number of changes had taken place, in addition to the building of the road.[38] Ranges A to E were still in use but a shorter range of 600yds (548m.) had been built to the west of range A and immediately adjoining it. Another shorter range of 500yds (457m.) had been laid out on the east side of range C and in close proximity to it. In addition a new range of 600yds had been built, midway between ranges C and D with the firing point on the west side of the road and adjacent to it. The firing points for all the other ranges were still located on the south-east side of the Tidworth road and presumably firing only took place at certain times on specified days when the road would be closed.

The Ordnance Survey map of 1957 shows that in the meantime, several changes had occurred. [39] The ranges had been reduced in length so that all the firing points were now on the left-hand side of the road when travelling in the direction of Tidworth. The effect of this was to reduce range A from 900yds to 400; Ranges B and C from 1,000 yds to 800; D from 900yds to 500 and E from 800yds to 400. At the same time the two short ranges of 600yds and 500yds which adjoined ranges A and C were removed while the 600yds range between C and D was closed. In 1970, ranges B and C were rebuilt to metric measurements and while B followed the line of the original range B, C was laid out a little distance to the west of the previous range.[40] The Ordnance plan which had been revised up to 1974, showed range A as 400yds, B and C each of 600m., D as 500yds and E 400yds.[41] Subsequently ranges A and E were increased in length from 400yds (365m.) to 400m. (437yds) and range D from 500yds (457m.) to 500m. (547yds). A sub-machine gun and pistol range of 100m. (109yds) designated range F, was also built to the north-east of range E.

Before the First World War, even rifle shooting had its lighter moments as can be seen from the following report which appeared in the *Andover Advertiser*.

> During the recent Territorial manoeuvres the subject of rifle shooting often cropped up.
>
> 'I'll bet anyone here a box of cigars' said Lieut. A, 'that I can fire 20 shots at 200 yards and tell without waiting for the marker, the result of each one correctly'.
>
> 'Done,' cried Major B. and the whole Mess turned out early next morning to witness the experiment.
>
> The lieutenant fired. 'Miss!' he announced calmly. Another shot 'Miss!'.
>
> 'Here hold on,' put in Major B. A third shot, 'Miss!'
>
> 'What are you trying to do? You're not firing for the target'.
>
> 'Of course not,' was the cool response, 'I'm firing for those cigars!'
>
> And he got them.[42]

As the Army made increasing use of the Plain two administrative problems arose. The first was the provision of adequate water supplies and in order to deal with the problem, it was agreed to spend £20,250 between 1903 and 1906

> for well-sinking and other work in connection with the supply of water at different parts of Salisbury Plain, necessary in order that more ground may be available for camping purposes.[43]

The second problem concerned the repair and maintenance of the roads,

> Many of the roads on Salisbury Plain have never been properly made up, as they were not extensively used until the establishment of the camps there within the last few years ... The Amesbury and Andover Rural District Councils having refused to bear any share of the cost of strengthening certain roads ... it was decided ... to strengthen these roads at the cost of Army Funds and then request the local councils to maintain them.[44]

Although the erection of the camp buildings was regarded by some as an intrusion into the landscape, the military presence gradually came to be accepted.

> As we pass above Milston with its little church ... the corrugated iron town of Bulford on the Plain beyond, comes painfully within range, and Sidbury Hill and other more distant heights about Tedworth ... seem to look even at this distance, as if worn by the tramp of horse and men and seared by the smoke of mimic warfare. But this is treason, for I can imagine no spot in all this island anything like so favourable for summering and exercising our hardly raised and none too numerous defenders. There are plenty of far larger moorlands in the north and west and in Wales, but they are not galloping grounds to put it mildly. Here we have a country dry and firm like the South African velt or the Manitoba prairies.[45]

In addition to providing for the construction of the road from Bulford to Tidworth the estimates for 1909–1910 also included the 're-appropriation' of F Lines as married quarters.[46] Other quarters were built in Vimy Crescent adjoining the south-west boundary of the barracks area in 1910 although this date cannot be confirmed. These consisted of five blocks each comprising twelve terrace houses of a similar design to those in Bazaar Road, Tidworth, and also known locally as 'Merthyr Tydfils'. At the same time two rows, one of five houses and the other of twenty, were built in Mons Avenue, adjacent to Wing Road. Other improvements continued to be carried out in the barracks including new accommodation for the Army Service Corps and sheds for the Royal Artillery.[47] After the electricity generating station had been built in Tidworth in 1913, a supply was brought to Bulford and came into use in 12th August 1914.[48]

One of the earliest references to the hospital in Bulford Camp appears in the Garrison Plumbers Work Book on 17th March 1902 so that it is evident that provision had been made for a hospital in the original plan of the barracks. The buildings were sited about 80yds (73m.) from and on the north side of, Horne Road access being provided by an entrance drive on the east side of the hospital.[49] In 1985, Kandy Road followed the line of this drive while the Police Station, numbers 6 to 9 Kandy Road and the NAAFI shop at the rear, occupied the site of the hospital. By January 1904 the 20th Company Royal Army Medical Corps, commanded by Lieut.-Col. C.E. Faunce, had moved to Bulford[50] and in April the *Salisbury Journal* reported that 'the number of beds in the Station Hospital at Bulford Camp is to be increased to 200 from Friday, 15th inst.'[51] This may have been related to the arrival of this unit but in the spring of 1908 it moved to Tidworth.[52] The hospital was still in use in 1921[53] but in 1929 and 1933 the respective *Barrack Books* only record it as a reception station. By 1939[54] all the buildings which were shown on the 1924 Ordnance map had been demolished. When the Australian Forces were stationed at Larkhill and Perham Down during the First World War, it was decided to establish an Australian hospital in Bulford. This was built in 1917[55] immediately to the west of the existing hospital and the entrance to it from Horne Road, has now become Bengal Road. It was considerably larger than the original Camp Hospital and the *Barrack Book 1921* shows that at that date, it contained 950 beds as compared to only 50 in the old hospital.

The Fifteenth Report of the Church of England Board for the Welfare of Soldiers for 1912, noted that 'the need of a Garrison Church at Bulford is very urgent, and the Board has decided to take this matter in hand at once'. During this period Bulford was considerably more important as an artillery station than Larkhill which only comprised the ranges and a number of sites for tented camps which were erected during the training season. Consequently the church was to have close Gunner connections from the beginning and at the first meeting of the Building Committee in 1913, Brig. F.D.V. Wing who had been a battery commander at Ladysmith, was appointed chairman.[56] As CRA of the 3rd Division his handling of the artillery at Le Cateau was described as 'masterly' and shortly afterwards he was promoted to command the 12th Division. Very sadly, he was killed on 2nd October 1915 at Fosse 7 about a mile and a half west of Loos[57] and Wing Road and Wing Barracks have been named in memory of him. Although the Bishop of Salisbury, the Rt. Rev. F.E. Ridgeway launched an appeal for funds early in 1914 and Messrs. Blount and Williamson of Salisbury were appointed architects, no further action was taken until after the end of the war.

In June 1920, the CRA, Brig.-Gen. L.C.L. Oldfield, CMG, DSO cut the first turf and on 13th October FM Lord Methuen, GCB, GCMG, GCVO laid the foundation stone in the presence of a distinguished company.[58] However, owing to a number of delaying factors, it was not until 23rd April 1923 that the work had advanced sufficiently for the nave to be dedicated by the Chaplain-General. When the church was finally completed, it was dedicated to St. George by the Bishop of Salisbury, the Rt. Rev. St. C.G.A. Donaldson, on 3rd November 1927.[59] Numerous gifts were made in respect of the fabric and fittings and these included the east window which was given by the Royal Artillery Memorials Society in memory of all ranks of the Royal Regiment who fell in the First World War.[60] The Pulpit was given by Mrs. Wing in memory of her husband, Maj.-Gen. F.D.V. Wing, CB.

To mark the fiftieth anniversary of its dedication, a special service was held in the church on 21st October 1973 which was attended by Gen. Sir Basil Eugster, KCB, KCVO, CBE, DSO, MC, Commander-in-Chief, United Kingdom Land Forces, Maj.-Gen. R. Lyon, OBE, Director Royal Artillery and Brig. H.L.B. Salmon Commander Salisbury Plain Area. The occasion was given a special significance by the presence of Miss Wing, daughter of Maj.-Gen. F.D.V. Wing.[61]

The story that lies behind the present Roman Catholic Garrison Church of Our Lady Queen of Peace, is far more complicated than in the case of St. George's. The first record of a Roman Catholic church at Bulford Camp relates to a church in G Lines later Bapaume Lines. This was in use from October 1910 until April 1912 but in June of that year another church was established in F Lines which was later named Messines Lines. In March 1915 this church closed and in June 1915 a third church was opened in G Lines where it remained until December 1926.[62] Although it is probable that one or possibly two of these churches were huts which were used for services when they became available, the Ordnance Survey map of 1925 (revised in 1923) marked and named two Roman Catholic churches. One of these was situated on the north side of Hubert Hamilton Road opposite G Lines and in 1985 the site formed part of the recreation ground next to the Sandhurst Block, being about twenty yards from the south-east corner of that building. The other church stood on what is now the car park opposite the Annexe Building of the Headquarters of South-West District in Rawlinson

Road and about eighty yards east of its junction with Signal Street.[63] This was in G Lines and was probably the last of the three churches mentioned above as it did not close until December 1926.

Exactly a year earlier, another church had been built on a site some 60yds (55m.) north of the present Catholic church and just to the north of the adjoining car park. This was a corrugated iron building which was dedicated to Our Lady of Victories and was in use from December 1925 until December 1968.[64] The present church of Our Lady Queen of Peace was completed in 1968 and was taken into use on 16th December of that year.[65] It was dedicated on 12th January 1969 and a commemorative inscription inside the east entrance, reads as follows.

THIS CHURCH WAS DEDICATED ON THE
12 JANUARY 1969 UNDER THE PATRONAGE OF
OUR LADY QUEEN OF PEACE
BY THE RIGHT REVEREND GERARD W. TICKLE
ROMAN CATHOLIC BISHOP IN ORDINARY
+ + TO HER MAJESTY'S FORCES + +

CHAPTER NINE

Bulford – The Scene Unfolds

With the completion of the new barracks, the need to provide for the welfare of the soldiers when they were off duty soon became apparent. The answer to the problem was supplied by three organizations: the Wesleyan Soldiers' Home, the Church of England Soldiers' Institute and Miss Perks' Soldiers' Home. Work began on the Wesleyan Home in 1902 and although it was opened in 1903 it was not completed until the following year.[1] The Home was built on a plot of land opposite the junction of Bulford Road and the Amesbury Road and adjoined the site on which St. George's Church was erected after the First War.[2] Unfortunately it was destroyed by fire in April 1922 and to replace it a building was purchased from the Congregational Church Authorities in December 1925.[3] This was already in use as a home since it is shown on the Ordnance map of 1924 and inscribed 'Institute'. It was known as the Cromwell Institute and apparently continued in use until after the end of the Second World War. In June 1985 it was still possible to read the words 'WESLEYAN SOLDIERS H...' which had been painted on the roof over the front entrance although the building was then in use as a Families Centre.

On 6th June 1903 the *Salisbury Journal* contained the following report.

> On Thursday the New Soldiers' Home at Bulford Camp was formally opened in the presence of a large company. The Home which has been built at a cost of nearly £7,000 is very pleasantly situated and is a commodious and handsome building ... Lieut.-Col. Stirling presided over the large attendance and he was supported on the platform by Lady Wimbourne who laid the Foundation stone in October last.[4]

Although no specific name is mentioned, the 'home' was the Church of England Soldiers' Institute and this is confirmed by Kelly's Directory which gives the Wesleyan Home, the Church of England Institute and Miss Perks' Soldiers Home as the only establishments of their kind at

Bulford Camp in 1903.[5] Miss Perks' Home is referred to later in this chapter. The Institute was built behind and adjoining the Garrison Theatre which stood on the south side of Plumer Road (now Wyvern Road) and to the east of St. George's church.[6] Although it was still in use in 1939 it was demolished after the Second World War and does not appear on the Ordnance map of 1961. However, the map of 1924[7] marks an 'institute' on the north side of Rawlinson Road, some 170 yards (155m.) east of Marlborough Road and on a postcard of the 1920s this is named the Church of England Soldiers' Home. In 1985 the building was still standing and was in use as a civilian club, known as the Gordon Club. The body responsible for establishing the original Institute was the Church of England Board for the Welfare of Soldiers details of which have already been given in Chapter 5. In 1908, the Board appointed a Bulford Institute management Committee under the chairmanship of Maj.-Gen. Sir Henry Rawlinson, Bt, CB, CVO (afterwards Gen. Lord Rawlinson, GCB, GCSI, GCVO, KCMG).[8] The Institute was clearly a success and the Report for 1911 refers to the building of accommodation for the staff, a stable and coach house.[9]

The third pre-1914 home was the one which had been established by Miss Perks and was known as 'Miss Perks' Soldiers' Home'; it is included in Kelly's Directory for 1903. Unfortunately very little information regarding Miss Perks has become available but in February 1904 she attended a reception given by the Mayor of Salisbury to mark the opening of a new building for the YMCA in Winchester Street. Speaking on this occasion she said that 'from her earliest childhood, she had been associated with YMCA work' and, referring to the value of the new building to the soldiers, she said that it was 'the only home they could go to in Salisbury and the only place they could go to outside Bulford Camp'.[10] The home occupied the site on which the present Roman Catholic Garrison Church now stands and the short avenue of lime trees leading to the Church originally lead to Miss Perks' Home.

The entry in *Kelly's Directory* for 1911 shows that the position regarding institutes at Bulford Camp was the same as it had been in 1903 and it was not until the beginning of the First World War, that the position changed. The rapid expansion of the Army after 4th August 1914 resulted in an unprecedented demand for canteen and recreational facilities which soon became a matter of considerable urgency. Amongst the voluntary organizations which came forward to meet this need were the Young Men's Christian Association ('The YMCA') and the Salvation Army. At Bulford, the YMCA provided four huts; the first of these was built on the north side of the school at the junction of Bond Street and Wing Road which in 1985 was in use as 'The Thrift Shop'. Another was erected in Sling Camp but the location of the two remaining huts has not been recorded.[11] The Second War saw renewed activity and about 1940 a YMCA building was erected next to the Cinema which stood opposite the junction of Haig Road and Marlborough

Road. This was generously presented by the Biscuit Trade Association and was affectionately known as 'The Biscuit Hut'; it was demolished in 1976. Next to this hut was a single storey brick building which was erected, probably in 1941, by the Young Women's Christian Association ('The YWCA'). The purpose of this was to provide some amenities for members of the Auxiliary Territorial Service (ATS), now the Women's Royal Army Corps. After the war it was given to the YMCA and was subsequently demolished with the main building in 1976.[12]

In 1914 the Salvation Army opened a hut in B Lines, between Haig Road and Grierson Road (now Wing Road). This was known as 'Bulford I Salvation Army Hut' and in 1915 another hut 'Bulford II' was opened 'near Miss Perks' Home' although the exact site has not been recorded.[13] In the same year 'Sling Salvation Army Hut' was opened in Sling Camp[24] and in March 1918, the first Salvation Army Hut for the Women's Auxiliary Army Corps (WAAC) which had been built behind Bulford I Hut, was opened.[15] No information is available as to when these huts were removed or what the position was between the Wars. By 1939 or early 1940, the Salvation Army were at Bulford once again and remained there until 1948 or thereabouts. In later years these canteens operated under the name of the Red Shield Canteens.[16]

In 1926 Miss Elise Sandes took over Miss Perks' Home which was then re-named the Sandes Soldiers' Home. Miss Sandes who was born in Tralee, Co. Kerry, in 1851 had been encouraged by a friend, Miss Maria Fry, to teach drummer boys in the Tralee Garrison, to read. Very soon her home became a meeting place for older soldiers as well as the boys, and in 1869 she rented a house which she adapted as a coffee room and rest centre and this became the first Sandes Home. By 1922 there were Homes in almost every garrison town in Ireland and a large number had been established overseas. After the partition, Miss Sandes moved her headquarters to the Ballykinler Home in Northern Ireland where she died in 1934 and was buried with full military honours in Tyrella Cemetery close to Ballykinler Camp. The Sandes Home at Bulford was closed in 1937.[17]

During the 1920s a branch of the Navy, Army and Air Force Institutes – the 'NAAFI' – which had replaced the Navy and Army Canteen Board on 1st January 1921, opened in what is now Ward Barracks. By 1939 a NAAFI Institute had been built in New Haig Road[18] and this is still in use. Unfortunately there are no records available from which information can be obtained regarding the work of the NAAFI at Bulford, during the early years. However, in 1985 the following NAAFI establishments were operating in the Garrison; the Junior Ranks Club in New Haig Road; the Flying Coachman in Horne Road; the White Horse Club between Rawlinson Road and Signal Street and the Families Shop in the Horne Road Shopping Centre. There was also a Messing Store in the old riding school building at the junction of Haig

Road and Mons Avenue while the NAAFI Regional Office occupied Cromwell House in Hubert Hamilton Road.[19] In 1974 the Saxon Warrior public house which is leased to Whitbread Wessex, was built on the north-east side of the car park in Kandy Road.[20]

Since the beginning of this century there have been at least six schools in Bulford Garrison at one time or another, the first three being located in Bond Street. At the south end of this street where it joins Bulford Road, was a small school which was probably built at the same time as the Barracks but by 1939 it had been demolished. The second school which stood at the junction of Bond Street and Grierson Road (now Wing Road) was the Elder Children's School which was apparently a Junior Mixed School. Old notebooks which belonged to the teaching staff show that the School was operating in January 1915 so that it had probably opened some years earlier.[21] In 1948 all the schools in Bulford were taken over by the Wiltshire County Council and in 1955 this school was named the Wing County Junior School as it was located in Wing Barracks.[22] However it was closed at the end of 1964 after the pupils had been transferred to the new Kiwi County Primary School and in 1985 the building was being used as 'The Thrift Shop'. The third was an infants school the corrugated iron buildings of which occupied a site between Bond Street, Haig Road, and Milne Road opposite the Wesleyan Soldiers' Home. The 1924 and 1961 editions of the Ordnance map mark this school, but on the map of 1970 the site is shown as a tennis court.

When, sometime after 1924, the massive lorry shelter at the end of Haig Road was removed, the area was developed by constructing Old Ward Road and building thirty-six houses around it. During the same period and before 1939, a school was built at the western end of this road and in 1968 it was considerably enlarged. This comprised a new building for 280 pupils erected on the south side of the old school which was incorporated in the new layout, while a nursery unit was added in 1974.[23] It is now known as Haig County Primary School. After the First World War and before 1939, an infants' school was built to the east of the junction of Marlborough Road and Hubert Hamilton Road on the site of the present Primary School. This was known as the Bulford Infants' School but was re-named the Kiwi Infants' School, probably after 1948.

By 1963 work had begun on the new Bulford Kiwi County Primary School and since the site that had been selected for it, was occupied by the Infants' School, the latter was closed in 1964 and demolished shortly afterwards. The new building was completed at the end of 1964 and came into use in January 1965[24] although the official opening by HE The Hon. Sir Thomas Macdonald, KCMG, the High Commissioner for New Zealand, did not take place until 6th July.[25] In 1985 a nursery school known as the Sling Playgroup School was operating at Club Road, Sling, in a building which had been part of Sling Camp during

(*HQ United Kingdom Land Forces*)

28. HRH The Duke of Edinburgh inspecting the 1st Battalion Queen's Own Highlanders (Seaforth and Camerons) at Tidworth on 23rd July 1983. With His Royal Highness is the Commanding Officer Lt.-Col. N. J. Ridley.

(Imperial War Museum)

29. A Review held by Gen. Sir Horace Smith-Dorrien at Perham Down on 22nd May 1913.

(*N. D. G. James*)

36. The Garrison Church of St. George (Church of England) Bulford which was completed in 1927.

(*N. D. G. James*)

37. One of the first schools in Bulford. The Elder Childrens School in Wing Road renamed the Wing County Junior School in 1955 but closed in 1964. In 1985 was 'The Thrift Shop'.

(*N. D. G. James*)

34. The 'Merthyr Tydfils' – the original married soldiers quarters in Vimy Crescent Bulford in 1985.

(*G. Ward*)

35. A Heavy Artillery Brigade camp at Bulford in 1915.

(*Ludgershall History Society*)

32. 25th Battalion The Royal Northumberland Fusiliers (now The Royal Regiment of Fusiliers) marching along Tidworth Road, Ludgershall during the First World War.

(*The Tank Museum*)

33. HRH Prince Henry, Duke of Gloucester in conversation with Lt. H. G. Neish who was commanding a medium tank of 5th Battalion, Royal Tank Corps, near Nether Wallop during the manoeuvres of 1926.

(*Imperial War Museum*)

30. A BE2a biplane of No. 4 Squadron, Royal Flying Corps, flying past at the Review at Perham Down. 22nd May 1913.

(*Ludgershall History Society*)

31. HM King George V leaving Ludgershall Station (where he had arrived by the Royal Train) for Perham Down on 8th November 1917.

(*N. D. G. James*)

38. First World War buildings in Club Road which were originally officers' quarters and then an officers' mess in Gallipoli Lines of Sling Camp. In 1985 the buildings were being used as a nursery school.

(*N. D. G. James*)

39. The Wesleyan Soldiers' Home in Bond Street Bulford. Originally known as the Cromwell Institute, in 1986 it was the Garrison Families Centre.

(L. Campbell)

40. The original Church of England Soldiers' Institute in Plumer Road (now Wyvern Road).

(G. Ward)

41. Miss Perks' Soldiers Home which stood on the site of the present Roman Catholic Church.

the First World War.

The first place of entertainment to be provided was the Garrison Theatre which was next to the Church of England Institute in Plumer Road. It is not clear for how long this functioned as a theatre and it appears that by 1939 it was used for showing films. There was also a cinema in Rawlinson Road which was situated between the old riding school (now the gymnasium) and the Gordon Club. However this had been demolished before 1939 and had been replaced by one built between the wars by A.N. Kendall who was also concerned with the Hippodrome Cinema in Tidworth. The cinema at Bulford was still standing in Marlborough Road opposite Haig Road in 1985, although it was then used as a 'Bingo Hall'.

Other buildings which played a prominent part in the life of the Garrison included the corrugated iron fire station which was still to be seen in 1985 at the junction of Rawlinson Road and Signal Street. The Garrison Market which consisted of an open fronted shed built of corrugated iron, in the shape of a hollow square, was situated between Bond Street and Mons Avenue and was similar to Tidworth market. At the northern end of Bond Street and close to the market were a number of shops which were of similar construction and consequently the area was generally known as 'Tin Town'. The market and shops were demolished in September 1969 and by 1985 the site of the market had become a car park. The large scale Ordnance map of 1924[26] shows a building which covered a larger surface area than any other that has been erected in the Garrison. Inscribed on the map as 'Lorry Shelter', it was situated at the western end of Haig Road and was approximately 207yds (189m.) long and 87yds (79m.) wide. The building covered the whole of the area on which the Old Ward Road housing estate and the Haigh Primary School now stand.

In October 1900 Bulford Camp was under the command of Col. (temporary Brig.-Gen.) V. Hutton, CB[27] and at first it was used to a large extent by cavalry regiments. During 1903 a School of Mounted Infantry was formed but when it was disbanded in the middle of 1906, the Cavalry School at Netheravon moved to Bulford as a temporary measure, and remained there until May 1907.[28] In January 1904, 20 Company RAMC was transferred to Bulford but in April 1908 it proceeded to Tidworth[29] as by that date a military hospital had been set up in Delhi Barracks. By 1905 a number of artillery brigades were stationed at Bulford since there was as yet no permanent accommodation at Larkhill. When the Cavalry School returned to Netheravon in 1907, its place was taken by the 3rd Division commanded by Maj.-Gen. W.E. Franklyn, CB. This Division which subsequently had a long connection with Bulford, remained there until it went overseas in 1914[30] but on its return after the war it was stationed at Bulford from 1920 to 1939.[31] It was disbanded in 1947 but was re-formed in 1959 – once again at Bulford which was described as 'its traditional home'.[32]

On 1st August 1977 it became the 3rd Armoured Division and was transferred to the British Army of the Rhine.[33]

Although the Tedworth Hounds had been kennelled at Tidworth for many years and had hunted the surrounding countryside, since they were foxhounds the Hunt had no interest in the large number of hares which were then to be found on the Plain. In 1905 several artillery brigades were stationed at Bulford and it was not long before some of the Gunner officers had the idea of getting some hounds together so as to hunt these hares. When in 1907, the Master of the Tedworth, Capt. A.E. Hussey, very kindly agreed to allow hare hunting in part of the country, it was decided to form a pack of harriers to be known as 'The Royal Artillery (Bulford) Harriers'. The original kennels were situated on the south side of Bulford Road about 100yds (91m.) from its junction with Milne Road (grid reference 182433). Owing to the First World War the pack unfortunately had to be put down in 1917 but it was re-established two years later. In 1934 the hounds were moved to new kennels which had been built between the railway line and the northern part of what is now New Ward Road. On the outbreak of war in 1939 hunting was stopped, but in 1942 the pack was re-formed as foxhounds under the title of 'The Royal Artillery Hunt' and it was still hunting regularly in 1985.

When war was declared on 4th August 1914, thousands of young men volunteered for the Services and some idea of what this new way of life meant to them, can be gathered from the letters which passed between Corporal Ralph Painter of the 4th Bn. The Bedfordshire Regiment and his fiancée. Writing from a canteen which he described as 'a nice place thanks to the YMCA' at Dovercourt near Harwich in October 1914, he said that although he had not yet been issued with any uniform, he had drawn his rifle. At the same time he confided that 'I have never had a shot with a gun in my life but when I do, I shall be all right' and he was obviously encouraged by the fact that 'we get better pay according to our shooting'. He had already attended several lectures on the use and care of the rifle in preparation for his first day on the ranges. Route marches played a large part in his training and he was given four shillings (20p.) by his company commander to buy mouth organs so as to provide some music for these occasions. He issued them to the men whom he thought were the best players before they moved off and collected them again on their return. After the first time that the mouth organs were played he wrote that 'it went off first rate'. He was obviously a keen young soldier and after describing his duties as Corporal of the Guard in detail added, 'I rather like it'. In a subsequent letter dated 17th October 1914 he proudly told his girl 'I am now in khaki'. It is interesting to read that although he had not been in the Army before – 'I am getting quite used to my new life' – and was completely untrained, he had been promoted to corporal within a few weeks of joining up. Without wishing to detract from his efficiency, it is probable that such early promotion was largely due to a shortage of trained NCOs brought about by the rapid expansion of the Army.[34]

At Bulford the outbreak of war was followed by the building of additional hutted camps the first of which was at Sling between Gaza Road and Sling Plantation. Work was begun in the autumn of 1914 by the British Section of the New Zealand Expeditionary Force, that is to say New Zealanders who were in the United Kingdom when war was declared. The camp was intended for the main body of the Expeditionary Force but when it was learnt that the Force had landed at Alexandria, the British Section was instructed to join them. The Canadian Expeditionary Force then took over and continued the work of building Sling Camp,[35] which was later occupied by the New Zealand Forces in 1916 when it became known as Anzac Camp.[36]

A panoramic drawing made by C.E.H. Pratt of the New Zealand Expeditionary Force during the War shows that Anzac Camp comprised four main sections: Auckland, Wellington, Otago and Canterbury Lines.[37] Auckland Lines were situated in the area between Sling Road and Baghdad Road, as the 'Sling by-pass' road had not yet been built. Wellington Lines which adjoined Auckland, occupied the land that in 1985 formed that part of Sling Officers' Married Quarters which lay between Baghdad Road and The Crescent. Otago Lines covered the area between The Crescent and the track at the foot of Beacon Hill, while Canterbury Lines were located about 500yds (457m.) to the east of Three Acre Plantation and on the south of the Tidworth road which formed the northern boundary. If the Ordnance map of 1924 which was revised in 1923,[38] is compared with this drawing, it will be seen that much of the camp was still in existence in 1923 although some parts had been demolished. The huts which remained were located in two areas; Gallipoli Lines on the west and Baghdad Lines on the east. It is probable that the original camp was finally demolished during the mid 1920s as *The Barrack Book 1929*, includes an entry under the heading of 'Sling Camp' which comprised nine officers' married quarters and eight for NCOs, all of which were built in 1925.[39]

In 1916 another block of huts was built midway between Hubert Hamilton Road and Sling Plantation, about 300yds (274m.) to the east of Marlborough road. After the war these were named Marne Lines and the 1929 and 1931 editions of *The Barrack Book* show that they were allocated to the 3rd Divisional Signals. When the Australian Forces Hospital was built in Horne Road in 1917, a number of huts were erected on the adjoining site between what is now Kashmir Road and New Ward Road, presumably for the hospital staff. This area was subsequently named Somme Lines.

The Canadian Expeditionary Force which was composed of Canadians and Newfoundlanders, was the first large contingent of Dominion troops to arrive on Salisbury Plain in the First World War.

They disembarked at Plymouth between 15th and 24th October 1914 and then proceeded to four tented camps on West Down near Larkhill[40] but no sooner had they arrived than the weather broke and an exceptionally wet period followed. Steps were taken to find better accommodation so that the troops could be moved out of their weather-beaten camps and on 9th November the 4th Canadian Infantry Brigade moved to Sling and took over the huts which had been completed.[41] However on 7th February the Canadian Expeditionary Force left the Plain for Avonmouth where it embarked for France.[42]

During the Gallipoli Campaign many New Zealand casualties were brought to England from the Dardanelles and by March 1916 there were some 2,000 men who were convalescent in this country. At first, after they had been passed fit, they were posted to an Australian depot at Weymouth but subsequently they were transferred to an exclusively New Zealand centre at Hornchurch.[43] As this was the only New Zealand camp in Britain at the time, it had to accommodate men who were fit as well as those who were convalescing. Not surprisingly this combination did not prove satisfactory and by mid-July arrangements had been made so that Hornchurch took those who were convalescing, those who were recovering were transferred to Codford, while those who were fit were posted to Sling.[44]

However Sling not only became a centre for re-training casualties who were fit again but also for training reinforcements for the New Zealand Forces in France and Flanders and was officially known as the 4th New Zealand Infantry Brigade Reserve Camp.[45]

> Sling was the temporary home of most of the men – Sling unloved, bleak and lonely. It remained the chief New Zealand training camp in England throughout the war ... Who will not remember the bullring of Sling, with its bare slippery surfaces, its bleak winds sweeping across, its snow covered rifle ranges or its dreary heat and its monotonous tasks?[46]

The 'bullring' appears to have been the area on the north side of the Tidworth road between Sheep Bridge on the west and the rifle ranges on the east. In spite of all this, Sling had some compensations.

> The huts were comfortable and warmed in winter, the food was wholesome, well cooked, well served on hot plates, the canteens well appointed and with good libraries and billiard tables; and the spacious YMCA with its devoted, kindly English ladies and its cinema, catered well for the leisure hours.[47]

On 31st October 1918, some ten days before the Armistice was signed, there were 162 officers and 4,167 soldiers in Sling Camp.[48] When the war ended the New Zealand units which were in France and Belgium, returned to England, the majority being posted either to Larkhill or Sling where they awaited repatriation. Unfortunately it was some time before the necessary shipping was available and in the meantime discipline and training continued as if the war was still on. In an atmosphere of anti-climax, frustration and impatience, it was hardly surprising that before long an outbreak of trouble occurred at Sling. After this had been dealt with, Maj. L.W. Metcalfe, MC, approached his commanding officer, Lieut.-Col. A.G. MacKenzie with the suggestion that instead of parades and exercises, a kiwi the emblem of New Zealand, should be cut out on the side of Beacon Hill just above the camp. The Colonel readily agreed and with the approval of Brig.-Gen. G.S. Richardson, CB, CMG the General Officer Commanding the New Zealand Expeditionary Force Establishments in the United Kingdom, Maj. Metcalfe visited the War Office and obtained permission to proceed.[49]

The first step was to make an accurate drawing of a kiwi to enable the outline to be plotted on the ground. This was undertaken by Sgt.-Maj. P.C. Blenkarne who was a drawing instructor in the New Zealand Army Educational Corps and in order to obtain the dimensions of a kiwi he visited the Natural History Museum in London. The design was then marked out on the site and when this had been done, digging began so as to expose the underlying chalk. When completed, the kiwi was 420ft. (128m.) long, its beak being 50ft. (46m.) and the adjoining letters 'NZ' 65ft. high (20m.), the whole covering about an acre and a half or a little over half a hectare.[50] Although Kiwi Products (UK) Ltd., had for many years generously contributed to the cost of keeping the chalk clear of weeds and making good any damage to the emblem, they were unable to do so after 1967. From 1968 to 1977 the New Zealand High Commission met the cost of upkeep but after 1977 the condition began to deteriorate.[51] However in 1980, 249 Signal Squadron (AMF (L)), Royal Signals commanded by Maj. J.H. Fisher, MBE (afterwards Lieut.-Col. J.H. Fisher) offered to undertake the complete restoration of the Kiwi. This offer was readily accepted and the work was carried out by the officers and soldiers of the Squadron who completed the work in three days, operating in four-hour shifts. As well as clearing the grass and bushes, nearly 100 tons of fresh chalk was excavated from a nearby quarry which had to be carried on to the site owing to difficulties of access. In August 1981, the New Zealand General Staff presented a carved model of a Kiwi to the Squadron in recognition of their work and this bears the following inscription.

THE KIWI TROPHY
PRESENTED TO
249 SIG SQN A.M.F. (L)
BY
THE C.G.S. N.Z. ARMY
FOR THE
ANNUAL BIATHLON COMPETITION

This competition which consists of shooting and physical endurance tests was first held on 8th September 1981 when the High Commissioner for New Zealand, HE The Hon. L. Gandar presented the trophy.

The Kiwi emblem is now regularly maintained twice a year by the Squadron.[52]

During the First World War the Australian Forces which were stationed on the Plain were principally at Tidworth, Perham Down and Larkhill but an Australian hospital, the 1st Australian Divisional Hospital was established at Bulford in 1917. This has already been referred to, earlier in this chapter. After the Armistice the Australian troops that were in France returned to England and by the time the last 10,000 arrived in May 1919, there were already 70,000 Australians on Salisbury Plain.[53] The majority of these were accommodated at Bulford and Tidworth where they remained until arrangements had been made for their return to Australia.

Sometime after 1914 an Interned Aliens Camp was established at Bulford and this is recorded in the War Office document concerned, under a section headed 'Accommodation of Prisoners of War'.[54] It is possibly on account of this that the camp which is thought to have been near Sling, is said to have been a prisoner of war camp.[55]

After the end of the War, Bulford returned to the more ordered life of a peacetime station but gradually changes began to take place. These were mainly due to the expansion of mechanization which was soon to alter the whole structure of the British Army and as a result some alterations were made to the barracks as for example, stables being replaced by vehicle standings. The first married officers' quarters to be built after the War were three detached houses which stand on the north side of Gaza Road near its junction with the Tidworth road.[56] In 1932 work began on the replacement of the original wooden huts with brick buildings but comparatively little had been done by September 1939 other than erecting a barrack block in Wing and two administrative blocks in Beacon Barracks.[57] On the other hand, a large number of married quarters were built during this period, some twenty-nine quarters for officers being erected in 1937 at Sling, in Baghdad Road, Gaza Road and on the east side of The Crescent.[58] In the same year, twenty-six married soldiers' quarters were built in Hubert Hamilton Road and Marne Crescent while in 1938 a further fifty-nine were erected in New Ward Road and thirty in Melbourne Avenue. Some of the houses in New Ward Road were used to accommodate members of the Auxiliary Territorial Service (now the Women's Royal Army Corps) during the War that followed.[59]

The last building project to be undertaken before the Second World War, was the construction of a new hutted camp which was named Carter Barracks presumably after Maj.-Gen. Sir Evan Carter, KCMG, CB, MVO. This had become necessary in view of the increase in the size of the Army which was due to several factors including the introduction of conscription in May 1939. Construction began early in 1939 and continued into 1940, the site being on the north side of the road from Bulford Village to Sheep Bridge and Sling with the entrance about 500yds (457m.) west of Sheep Bridge. The area which lay between the new barracks and Bulford Village had been a tented camp site for a number of years. Carter Barracks were demolished in 1978 and in 1985 the old system of roads which had served the barrack area were in use as a drivers' training area administered by Headquarters, South West District.

Until 1922 Bulford Garrison was divided into eight 'lines', seven of which were lettered A to G while the eighth was known as the ASC Lines since they were occupied by the Army Service Corps. On the west side of Marlborough Road the lines were located as follows: The ASC Lines between Horne and Haig Roads and west of Bond Street; A Lines between Bulford Road and Grierson Road (now Wing Road); B Lines between Grierson Road and Haig Road and C Lines between Haig Road and Horne Road. To the east of Marlborough Road D Lines lay between Plumer Road (now Wyvern Road), Rawlinson Road and Veterinary Road which at that time was the southern extension of Signal Street while E Lines lay between Rawlinson Road, Hubert Hamilton Road and Signal Street. F and G Lines were extensions of D and E respectively, their eastern limits being Headquarters Road and the prolongation of that road northwards.

In 1922 names were given to each of the lines in commemoration of battles of the First World War, the ASC Lines becoming Amiens Lines and Lines A to G being designated respectively Mons, Cambrai, Arras, Ypres, Aisne, Messines and Bapaume. At the same time the hospital area on the north side of Horne Road was named Somme Lines and a block of huts situated between Sling Plantation and Hubert Hamilton Road was given the name Marne Lines. Sling Camp, some sections of which had been demolished by 1923 was divided into two sections; part of the area which lay between the western length of Baghdad Road and Sling Road being designated Gallipoli Lines, while the area to the east of Baghdad Road became Baghdad Lines. However this system was discontinued as from 1st April 1938 and was replaced by named barracks which were composed of one or more lines.[60] The names that were allocated to these were Beacon, Gordon, Kiwi, Marlborough, Ward and Wing but by 1961 Beacon Barracks had been re-named Picton, while Marlborough Barracks had ceased to exist since part of it had been included in Ward and part in Sling. A summary of the changes which have taken place since 1902, will be found in Appendix VII.

During the Second World War, Bulford was mainly occupied by British formations but some Commonwealth troops were stationed there during the latter half of 1940. From June to December of that year a unit of the Royal Australian Army Ordnance Corps occupied the set of buildings which are on the west side of the Amesbury road and 125yds (115m.) south of its junction with Bulford Road.[61] These have been used by the Royal Army Ordnance Corps for many years and

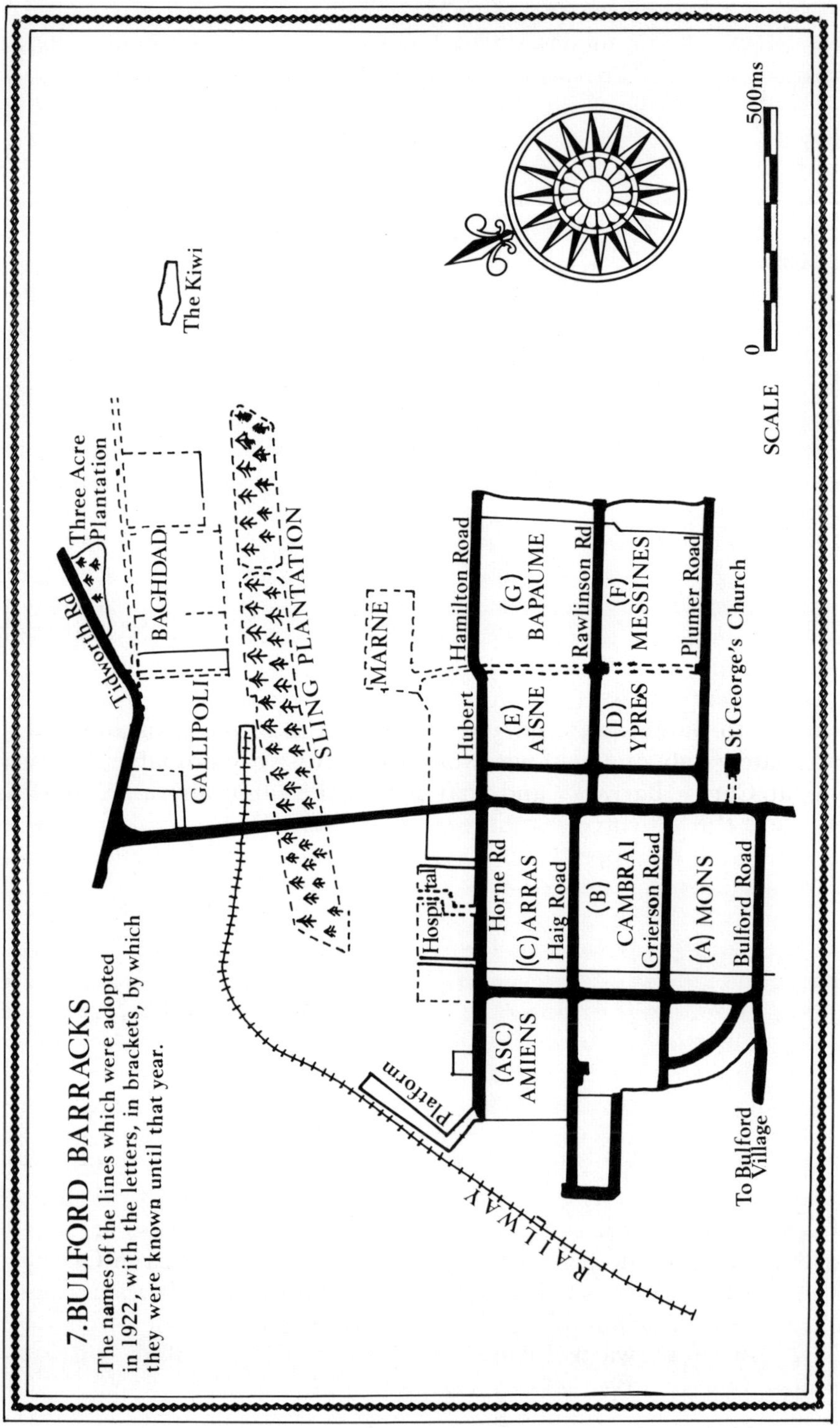

7. BULFORD BARRACKS
The names of the lines which were adopted in 1922, with the letters, in brackets, by which they were known until that year.

during the 1920s were known as Mobilization Stores but in 1985 they formed the Accommodation Exchange Store. A detachment of the Australian Army Provost Corps was also stationed in Church Road (now Provost Road) near the site of the present headquarters of the Royal Military Police[62] while other Australian troops were under canvas on the camp site next to Carter Barracks.

One afternoon in July 1940, Mrs. J. Philpots who was then a young girl was at Bulford Station with her dog, checking the number of railway wagons of coal which had been invoiced to her father who ran a coal business in Bulford. The trucks were standing in a siding next to several others which were being unloaded by Australian soldiers. Suddenly there was the sound of an aircraft and a low flying plane was seen approaching the station from the south-west. Immediately one of the Australians seized Mrs. Philpots by the hand and pulled her and her dog under one of the trucks and at the same time the station yard was heavily machine gunned, some of the bullets striking the rails on which the trucks were standing. However, thanks to the very prompt action of the Australian soldier she and her dog escaped injury.[63] A history of the Australian Imperial Forces during the Second World War entitled *To Benghasi* contains a reference to what may have been the same occasion which took place on Salisbury Plain although the exact location is not stated. On 13th July 1940 'a German bomber appeared out of low cloud and as it passed over the lines of the 2/9th and 2/10th Battalions, fired 300 to 400 rounds at the tents and huts'; fortunately only one man was wounded.[64] It is very probable that this was at Carter Barracks and that it was the same aircraft that had attacked the railway wagons, since the Barracks were in line with the station when approached from the south-west and were less than 1,500yds (1,370m.) beyond it.

There was also an Indian Mule Company of the Royal Indian Army Service Corps at Bulford in June 1940[65] which camped on the north side of the sports ground between New Ward Road and Sling Road and in the adjoining woodland.[66] Brigadier F. Fitzgibbon, DSO referred to a small detachment of Indian troops which he saw at Amesbury station after Dunkirk and it may have been part of the Mule Company.[67]

During the course of the War, Bulford saw the arrival and departure of many units and formations and of these, one was unique in that it presented an entirely new approach to operations which until then, have been adopted by the British Army. In March 1942 the 1st Parachute Brigade which formed part of the 1st Airborne Division, moved to Bulford where it remained until it left for North Africa in October of that year.[68] The Plain provided ample facilities for airborne training and the 1st Parachute Battalion under the command of Lieut.-Col. E.E. Down (afterwards Lieut.-Gen. Sir Ernest Down, KBE, CB) was stationed in Kiwi Barracks.

I remember that Kiwi Barracks were composed of relatively modern buildings, comfortable and well equipped. We had the use of a gymnasium, parade ground, sportsfields and the NAAFI, quite a significant change from the facilities available to us at the 'scratch' accommodation that we had previously experienced ... The foundations of the Airborne Corps, indeed, were laid at Bulford and on Salisbury Plain, and those of us who survived remember our time there with gratitude and some affection.[69]

In the meantime, the formation of the 6th Airborne Division, like that of the 1st Division, had been planned at Syrencote House some two miles north-west of Bulford Camp. These events are recorded on a plaque which has been placed beside the front entrance of the house and is inscribed as follows

Syrencote House
This house was used between November 1941 and
May 1943 as the headquarters of Major-General, later
Lieutenant-General F.M. (Boy) Browning, CB, DSO, the founder of
Airborne Forces. Much of the planning which lead to the
formation of the 1st and 6th Airborne Divisions was done in this house.
In May 1943, it became the headquarters of the
Sixth Airborne Division commanded by Major-General R.N. Gale, OBE, MC
later General Sir Richard Gale, GCB, KBE, DSO, MC
The Airborne Assault by the 6th Airborne Division on the night of
5/6 June 1944, to secure the left flank of the Allied Invasion of
Europe – Operation Overlord was planned and mounted from this house.
AD UNUM OMNES.

The post-war years were marked by extensive building operations in the Garrison which began in the 1950s but reached their climax towards the end of the following decade. In 1952 additional married officers' quarters were built at Sling more particularly in Auckland Road, Wellington Road, the west side of The Crescent, Kiwi Road and Beacon Hill Road. This was followed by an interval of several years except for the building of three married quarters in Milston Road in 1954, but in 1963 there was renewed activity. During that year, a large number of married soldiers' quarters were erected at the eastern end of Hubert Hamilton Road and on the adjoining land which is now known as the Australian Estate on account of the names of the roads – Adelaide Close, Brisbane Gardens, Canberra Drive and Darwin Close.[70]

From 1967 to 1976, the rebuilding of Picton Barracks, which in 1961 was known as Beacon Barracks,[71] was carried out. One of the results of this was the elimination of three roads which connected Rawlinson Road with Plumer Road, namely, Veterinary Road and the southern sections of Sapper Street and Church Street (now Provost Road). During 1968 further quarters for married officers were built at Sling in Nelson Close while a large number of married soldiers' quarters were erected on the north-east side of Bulford Road between Vimy Crescent and the line of the old railway. Since the names of the roads in this area have a connection with Canada and Newfoundland it is frequently referred to as the Canadian Estate. At the same time work began on the Irish Estate to the west of New Ward Road which provided married quarters for soldiers in Antrim Close, Londonderry Gardens and Tyrone Avenue. In 1969 quarters were also built in Wiltshire Close and Dorset Close to the west of the Canadian Estate and on the opposite side of Bulford Road while in 1976 the re-building of Gordon Barracks was completed.[72]

While the construction of the new barracks and married quarters provided the clearest evidence of the changes that were taking place during the post-war period, there were also others. In June 1952, passenger trains were withdrawn from the branch that ran from the Salisbury main line to Amesbury and Bulford while in March 1963, goods traffic ceased. The end of the line came in June 1965 when work on lifting the track began. On 18th September 1969, the new Salisbury Plain Area Works Office of the Property Services Agency which had been built on the site of the station in Bulford village, was opened. To mark its connection with the railway, one of the old signals from the line was erected at the beginning of the approach road to the office.

In 1977 the Headquarters of South West District was transferred from Taunton to Bulford and to mark the occasion Plumer Road was re-named Wyvern Road on 31st July of that year. The history of South West District can be traced to the Plymouth Garrison in 1803 which became the Western Garrison in 1810 and since then it has operated under a succession of titles. From 1835 to 1923, it was known as the Western District; from 1924 to 1937 as the Wessex Area and then, for two brief periods of two years, as the South West Area and the Western Area. Finally, in 1943 it became the South West District except for five years from 1961 to 1966, when it was named the 43 (Wessex) District.

CHAPTER TEN

Larkhill and West Down

During the latter half of the 19th century the need for permanent artillery ranges of adequate size, soon became apparent. There were three main reasons for this: the adoption of new principles in the design of guns, the introduction of new types of explosives and changes in the tactical employment of artillery which required practice in fire and movement.

In 1854, a Tyneside engineer, William Armstrong (afterwards Lord Armstrong) produced a breech loading rifled gun which incorporated an entirely new concept in its construction. Instead of using cast iron, the barrel consisted of a steel tube around which layers of wrought iron were shrunk thus giving it greatly increased strength.[1] Although this design was accepted for issue to horse and field artillery, by 1865 a body of opinion had emerged which favoured a return to muzzle loading guns and a committee appointed to consider the matter recommended their retention.[2] In the following year another committee reached the same conclusion[3] but by 1885 the situation had changed.[4] This was largely due to the replacement of gunpowder by more powerful propellants which required stronger barrels and also to the fact that improvements had been made in the design of the breech mechanism. By 1892, cordite had been adopted as a propellant[5] so that heavier shells could be used and gun ranges increased.

When the School of Gunnery was established at Shoeburyness on 1st April 1859, it was laid down that it should provide 'for the instruction of Horse, Field, Mountain and Garrison Artillery'.[6] However, while horse, field and mountain batteries could fire out to sea from a predetermined gun position, it was not possible to combine fire with movement as the necessary land was not available.[7] So as to overcome this problem a practice camp and range was set up at Okehampton in Devon some twenty-three miles west of Exeter which came into operation on 2nd August 1875[8] and was still in use in 1985. While the new range was of the greatest value it had certain disadvantages: parts were very steep or covered with scattered boulders which made the movement of horse batteries difficult; local mists could delay firing while Okehampton was some distance from the principal military centres in the south of England.

In some parts of the country, limited artillery practice was occasionally carried out, often by volunteers, on any open land which, with the consent of the local landowners, was available. It is probable that safety precautions were not of a very high standard and now and again rounds would 'go in the wrong direction' or 'go a bit too far'. This would result in alarm, verging in panic, in any villages or hamlets which received any stray rounds. However in 1877 three more 'Camps for Artillery Instruction' were established at Aldershot, Hay and Romney where a certain amount of firing practice could take place.[9]

Although the War Office Salisbury Plain Committee which was responsible for the acquisition of land on the Plain, held its first meeting on 11th May 1897, it was not until 20th April 1898 that any reference to an artillery range appeared in the minutes. It was then recorded that the proposed range was to be 'on the west side of the Avon' and the question of closing roads and tracks was discussed at some length. It was further stated that

> It was not intended that this proposed range should take the place of Okehampton, but it would take off a good deal of the pressure and would be an infinitely superior practice range to Shoebury.[10]

A year later the Committee agreed that the range area could be used for firing practice for the first time, from 14th to 24th August 1899.[11] The boundaries of the original range are marked on an Ordnance map revised in 1899 and published in 1903. This shows the western boundary following the old road from Rollestone cross roads to Redhorn Hill, then bearing eastwards to Casterley Camp where it turned south to meet the Packway at its junction with Tombs Road and so along the Packway back to Rollestone.[12]

By the end of the summer of 1899, it was realized that the area was too small and at a meeting on 26th October, the Committee approved that the range should be extended.[13] While there is no further reference to this in the Committee's minutes, the Annual Report of the School of Gunnery for 1901 stated that 'The extension of the artillery ranges to the westward as far as the Salisbury–Devizes road has been a marked success'.[14] It is not clear what additional land was included in this extension but the School of Gunnery's Report for 1911 noted that 'A new range at West Down was opened this season and should, when extended as proposed, form a valuable addition to the few acres available for gun practice'.[15] An Ordnance map published in 1906 on which revisions were overprinted in 1911 and 1914 marked two ranges – 'Larkhill Artillery Range' with the boundaries as already described, and 'Rollestone Artillery Range' now known as West Down range.[16] The eastern boundary adjoined Larkhill range while the southern, western and northern boundaries followed a line which ran approximately from The Bustard to Orcheston, Tilshead, Brazen Bottom Farm, Urchfont Hill and Redhorn Hill.

Although Bulford was only a few miles away, Gunner brigades which were stationed there did not normally practice at Larkhill or West Down. Mr. E. Hall who served in the 25th Brigade RFA at Bulford before the First World War, confirmed that his Brigade always went to Okehampton or Trawsfynydd some five miles south of Ffestiniog in North Wales, for practice camps. Apart from Territorial units, Larkhill was used, to a large extent, by brigades from Aldershot which marched there and back and as the journey took two days, one night was spent at Popham some seven miles south-west of Basingstoke.[17] Measured in a straight line, the distance from Aldershot to Popham was twenty miles and a further twenty-six miles on to Larkhill. It is possible that the location of the staging point was changed at some date, since in April 1904 the *Salisbury Journal* contained the following report.

> In connection with the route marching of the artillery of the First Army Corps from Aldershot to Larkhill Camp to undergo their annual course of gun practice on Salisbury Plain, a rest camp has been established off the London road at Andover where the troops going hither and thither will encamp for one night.[18]

Until 1914 there were no permanent buildings at Larkhill and units which attended for practice were accommodated in tents erected on official camp sites. The first three to be established were the 'RFA Practice Camp' at Larkhill, West Down North Camp and West Down South Camp and each is shown on an Ordnance map of the ranges which was published in 1903.[19] The RFA Camp was situated in the area which now lies between the RA Mess, Bell Road and the Garrison Church. The North Camp was located about 2,900yds (2,650m.) north-east of the present West Down Camp at Tilshead while the South Camp was some 2,500yds (2,285m.) east of it. From 1904 to 1906 the North and South Camps were also used by cavalry regiments so that they could practice with the 3.7mm. Maxim heavy machine guns, popularly known as 'pom-poms' with which they had recently been equipped.[20]

A similar map which was issued in 1906, recorded a number of changes that had taken place in the mean time. On West Down, a new camp known as Pond Farm Camp, had been formed some two miles to the north of West Down North Camp and about 650yds (595m.) west of the old road from The Bustard to Redhorn Hill. This must have been the most isolated camp on Salisbury Plain. The RFA Practice Camp was now re-named Durrington Camp while some 700yds (640m.) to the west, a new camp, named Lark Hill Camp, had been laid out on both sides of the Packway. Further west was Fargo Down Camp which was situated between the Packway and the northern end of Fargo Plantation. Immediately to the north of Rollestone Bake Farm was the original Rollestone Camp while opposite The Bustard and occupying both sides of the road was Bustard Camp. A later map based on that of 1903 but with revisions by the War Department up to 1914 showed all the foregoing camps although some had increased in size. One new camp – Hamilton Camp – named after Gen. Sir Ian Hamilton, GCB, GCMG, DSO was shown on the north side of The Packway, opposite Fargo Down Camp. At a later date an ammunition depot was built at the western end of Hamilton Camp but this was subsequently replaced by a larger depot on the south side of the road.

By 1910 a definite pattern for training during the summer months had been evolved and the *Andover Advertiser* reported

> everything points to a particularly busy training season on Salisbury Plain this year for the Regular troops, Special Reserve and Territorials. The canvas season will open in May and continue without intermission until the middle of September. It is pretty certain that on ... the western side of the Avon there is likely to be canvas centres at Larkhill, The Bustard, North and South West Down and Rollestone.[21]

One of the results of the arrival of the Army on the Plain was the decision to revive The Bustard Inn which stands on the edge of the ranges rather less than a mile to the north of the present-day Rollestone Camp. During the course of years the licence had been allowed to lapse but in February 1910, steps were taken to restore it.

> An application is to be made at the Salisbury County Petty Sessions next Tuesday for a licence to resuscitate The Bustard on Salisbury Plain, as licenced premises. As is well known The Bustard is situated in the very heart of the western territory of the Plain, several miles from anywhere and years ago it was the hostelry of The Bustard Inn ... The establishment of an hotel at The Bustard would be a great boon to Pressmen and other civilians whose avocations call them to the Plain during the summer.[22]

This application was successful and a new licence was granted some two months later.[23]

For several years preceding the First World War, the headquarters of the School of Gunnery for Horse and Field Artillery at Shoeburyness had moved to Okehampton each summer when batteries were attending practice camp. As the advantages of Larkhill became apparent it was urged that the time had come to transfer the School from Shoebury to the Plain. Although an opportunity to move it to Netheravon had occurred in 1900, it was rejected mainly on the grounds that the proposed accommodation was unsuitable and that it was too far from the ranges.[24] However, in the summer of 1912 the School was attached to Larkhill instead of to Okehampton and in 1914 it was finally decided that it should be moved to Larkhill. Subsequently an Army Council Instruction was issued in February 1915 which included the following.

> It has been decided to establish, as a temporary measure, a new school of instruction for R.H. and R.F.A. near Larkhill, Salisbury Plain. The new school will be on the lines of that already existing at Shoeburyness...[25]

The Instruction also directed that accommodation for the School was to be 'in or near Camp No. 14, south of Knighton Down'. This camp which was one of the thirty-four which are mentioned below was situated between what is now Brind Road, Glover Road and Knighton Down and forms part of Alanbrooke Barracks.

Until the end of 1914, very few changes had taken place at Larkhill since the land had been acquired by the War Department. On the ranges 'splinter proofs' had been built and 'danger posts' erected, tented camp sites had been laid out and water provided but unlike Tidworth and, to a lesser extent, Bulford no building development had taken place. Brig. F.E. Spencer, DSO, MC has recorded his impression of Larkhill as it was at the beginning of the First War when he was posted as a captain to 33rd Brigade RFA which was stationed at Bulford.

> In September 1914 we moved to Larkhill to carry out practice and at that time it consisted of a hut for the Gunnery Instructor on Knighton Down, all personnel being accommodated under canvas. From that single hut on Knighton Down has developed the great establishment which is now in existence there.[26]

However, after war was declared, the position changed dramatically when Larkhill was chosen as the site for one of the large hutted camps which were being built in different parts of the country.

Having decided on its location, the next step was to provide the means for moving the materials and personnel required for the construction of the camp. At the same time there was the problem of transporting troops, rations, fodder, ammunition, stores and so on when the camp was completed. Since motor transport was still in the early stages of development, road haulage was largely dependent on horses or steam traction engines such as those which were operated by Messrs. W.E. Chivers & Sons of Devizes. Clearly these methods of transport could not meet the needs of the new camp and it was consequently decided to build a railway from Ratfyn, some 700yds (640m.) north of Amesbury station to Larkhill. From there it was extended through Fargo Plantation to Druid's Lodge on the Salisbury–Devizes road (A360) with a branch to Stonehenge Aerodrome and another to Rollestone Camp. Known as the Larkhill Military Railway, it was built in 1914–1915 by one of the largest civil engineering contractors at the time – Sir John Jackson – who also erected all the hutted camps at Larkhill.

These numbered thirty-four in all which were completed in 1915[27] and were distributed as follows: numbers 1–14 lay to the east of a line from the present RA Mess southwards through the site of the old Down Barn destructor to Fargo Road. Numbers 15–30 were located to the

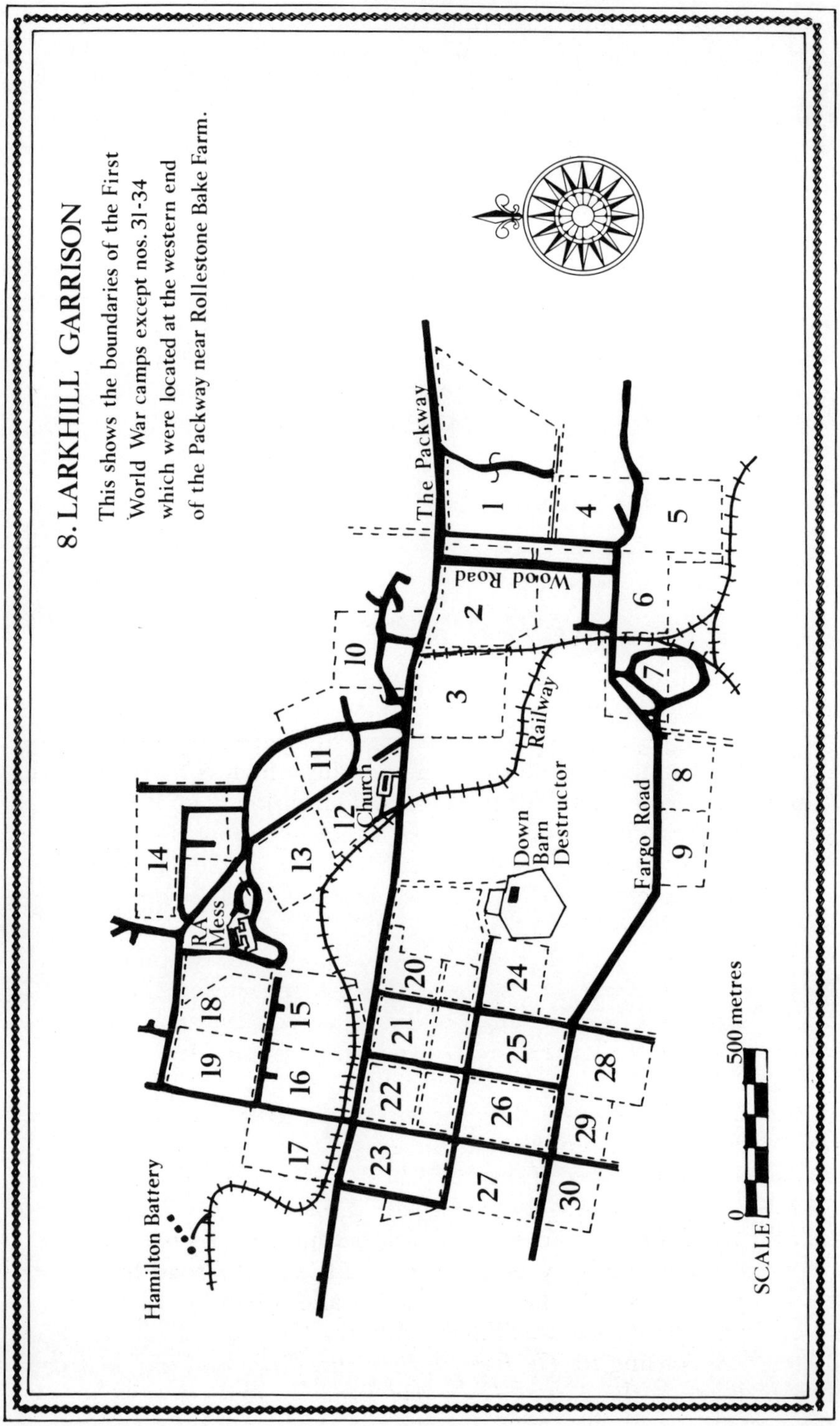

8. LARKHILL GARRISON

This shows the boundaries of the First World War camps except nos. 31-34 which were located at the western end of the Packway near Rollestone Bake Farm.

west of that line (numbers 15–19 lying to the north of the Packway and the remainder to the south) while numbers 31–34 formed a detached group about 545yds (500m.) east of Rollestone cross roads.

> In 1917 Larkhill consisted of wooden and tin hutted camps – some huts on brick supports about two to three feet high while all the horse lines and gun parks were in the open.[28]

Although Larkhill has always been associated with the Royal Artillery, during the First War the new hutted camps were occupied by 'units of practically every arm of the British Army'[29] each camp being designed to accommodate an infantry battalion. The emphasis on infantry training is reflected in some verses entitled 'Lark Hill Camp' which were written in June 1916 and included the following lines.

> There's an isolated desolated spot I'd like to mention
> Where all you hear is 'Stand at ease', 'Slope arms', 'Quick march', 'Attention'
> 'Slope arms', 'Fix bayonets', then 'Present', they don't half put you through it,
> And as you stagger to your hut, the Sergeant shouts 'Jump to it'.[30]

There was also a School of Musketry at Larkhill with rifle ranges at Rollestone and Durrington which are referred to in the Reconstruction Committee's report of 1918.[31] The ranges at Rollestone were situated between the Camp and the site of a tumulus which is marked on the Ordnance map of 1926[32] near Warren Plantation and Robin Hood Ball. There were four ranges, each of which were 600yds (548m.) in length with the firing points a little to the north of the Camp and the butts approximately in line with the tumulus. On the Durrington ranges there were five butts sited at distances of between 700yds (640m.) and 1,000yds (914m.) north of, and roughly parallel to Martinbushes Road. These were not, however, the first rifle ranges to be laid out near Larkhill since an Ordnance map of the area, published in 1903, shows that there were ranges at Honey Down Barn, Rushall Down, Shrewton Folly and Wilsford Down, details of which are given in Appendix VI. It is very unlikely that any of these ranges were in operation in the First War as they are not included in the *List of Lands and Buildings* which was issued in June 1918. If they had been in use during the early part of the War, it is unlikely that they would have been closed at a time when the need for rifle ranges could seldom have been greater.

Approximately midway between Larkhill and the road from Long Barrow cross roads to Rollestone (B3086) and adjoining the northern end of Fargo Plantation was Fargo Military Hospital which was built in 1914.[33] According to *The Barrack Book 1921*[34] the hospital contained 1,200 beds while the *List of Lands and Buildings* also refers to accommodation for the nursing staff. During the 1920s some of the buildings were used as quarters for RAF officers who were stationed at Rollestone and by 1929 these had been re-named Fargo Lodge,[35] while the Ordnance map of 1939 shows that in the meantime, many of the buildings had been demolished.[36] The remaining buildings had been removed by 1963 and in 1985 the site was marked by a scattered group of trees. Between the original Rollestone Camp and the Packway and some 800yds (730m.) east of Rollestone cross roads was the original ammunition compound which was served by a branch of the Larkhill Military Railway. Sometime after 1926 the compound was moved to its present position south of the Packway and about 220yds (200m.) nearer to Larkhill. There was also a prisoner of war camp which appears to have been located in the area to the east of Wood Road.[37]

As the importance of flying began to be appreciated from a military point of view, Larkhill became one of the main centres for army aviation from 1909 to 1912. In 1986 the original hangars built by the Bristol Flying School in Wood Road during this period, were still standing and in use as Accommodation Stores. During the First World War, two Kite Balloon Schools were formed at Rollestone and Tilshead respectively[38] and after the War the Rollestone School continued as an RAF station until December 1940. Further information on this will be found in Chapter 12.

In 1916 an artillery school, at first known as the Overseas Artillery School and later re-named the Chapperton Down Artillery School, was set up with its headquarters at 34–36, Fisherton Street, Salisbury. The staff and those attending courses at the School were billeted in several Salisbury hotels including the Old George Inn and The White Horse. The purpose of the School was to standardize gunnery methods and terminology in artillery formations from overseas, and demonstration shoots took place regularly on Chapperton Down which is now part of the Imber ranges. When the School of Artillery was formed at Larkhill in 1920 the Chapperton Down School became part of it.[39]

In the middle of October 1914, the first Canadian and Newfoundland troops arrived at Plymouth and after disembarking moved to four camps on or near West Down. The headquarters of the 1st Canadian Division were established at The Bustard while the 1st Canadian Infantry Brigade, the Divisional Mounted Troops and Princess Patricia's Canadian Light Infantry were allocated the adjoining Bustard Camp. The 2nd and 3rd Infantry Brigades occupied West Down South with the whole of the Divisional Artillery and Divisional Supply Column in West Down North. The remaining Divisional troops which included the 4th Brigade, the cavalry, the 17th Battalion and the Newfoundland Contingent, occupied Pond Farm.[40] The units bound for this camp detrained at Patney Station approximately midway between Westbury and Hungerford, shortly after midnight on 21st October and then marched to Pond Farm which they reached at three o'clock in the morning.[41]

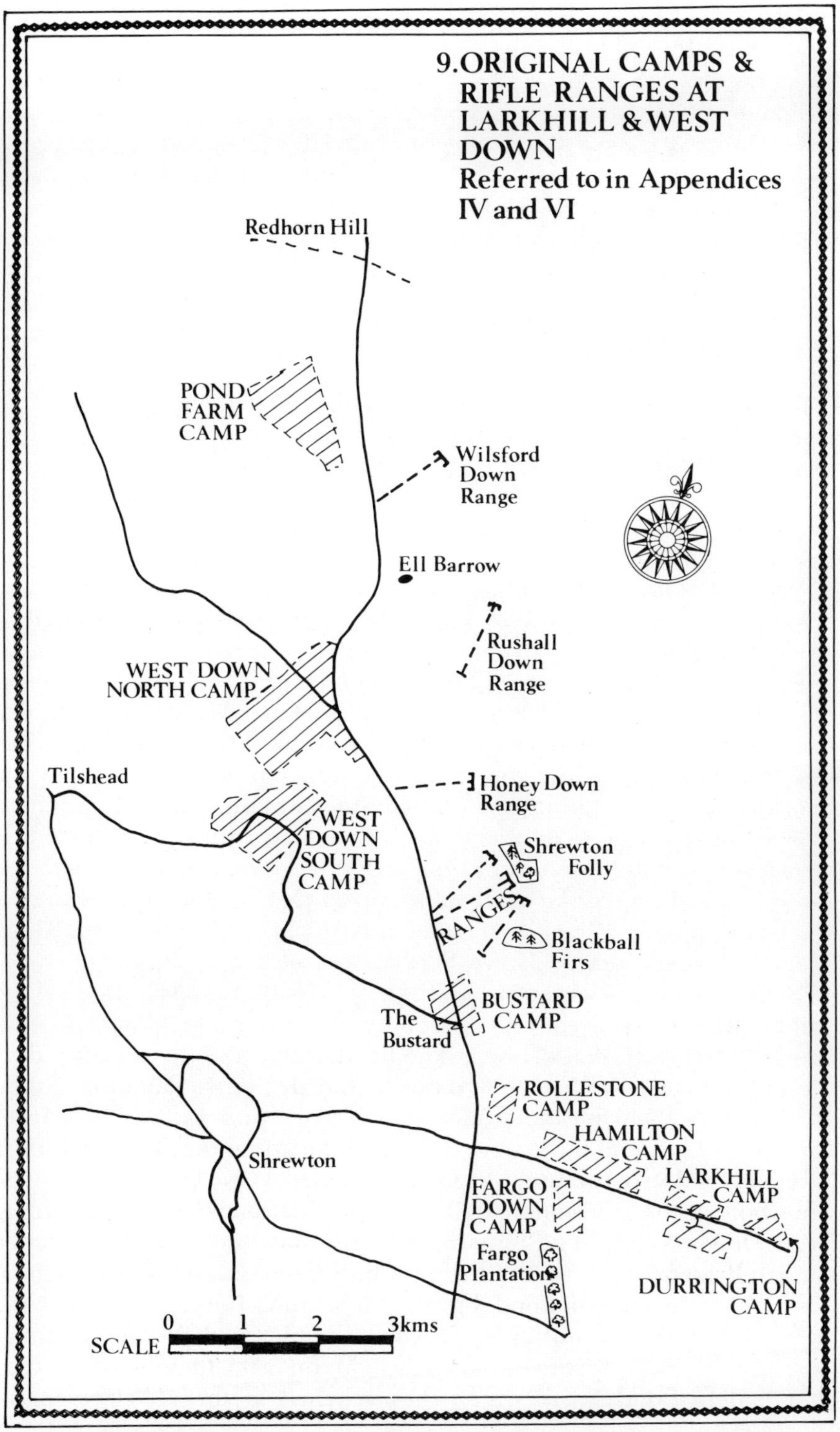
9. ORIGINAL CAMPS & RIFLE RANGES AT LARKHILL & WEST DOWN
Referred to in Appendices IV and VI
Redhorn Hill
POND FARM CAMP
Wilsford Down Range
Ell Barrow
Rushall Down Range
WEST DOWN NORTH CAMP
Tilshead
Honey Down Range
WEST DOWN SOUTH CAMP
Shrewton Folly
RANGES
Blackball Firs
BUSTARD CAMP
The Bustard
ROLLESTONE CAMP
HAMILTON CAMP
Shrewton
LARKHILL CAMP
FARGO DOWN CAMP
Fargo Plantation
DURRINGTON CAMP
0
1
2
3kms
SCALE

Although during September the weather had been warm and dry, by the middle of October the fine weather came to an abrupt end and was followed by an exceptionally wet period. This coincided with the arrival of the Division on 21st October and from then until the middle of February 1915, nearly 24 inches of rain fell[42] or an average of about one inch every five days. It was not only extremely wet but it was also cold with high winds to which the unheated tents offered little protection while the camp sites soon became a sea of mud.

> The miserable weather turned training into a drudgery. There were no means of drying clothing and men who ploughed through ankle-deep mud all day, had to let their rain soaked uniforms, dry on their backs.[43]

The first steps which were taken to deal with this problem, were to provide as much hutted accommodation as possible in place of tents. The British Section of the New Zealand Expeditionary Force had already begun the erection of huts at Sling but before the work had proceeded very far, the Section had been posted to Alexandria to join the main Force.[44] After the New Zealanders had gone, carpenters, bricklayers and unskilled labour were lent by the Canadians in order to help the civilian contractors complete the work at Sling.[45] Similar action was taken at Larkhill where Newfoundland and Canadian troops who had experience of this kind of work, were sent so as to speed the construction of the Camp.[46]

On 9th November, the 4th Infantry Brigade left Pond Farm and moved to the hutted camp at Sling[47] while at the end of that month, Princess Patricia's Light Infantry left Bustard Camp and proceeded to France with the 27th British Division.[48] On their departure, the Newfoundland Contingent took over their tents at The Bustard but this was only to be a temporary arrangement as on 7th December the Newfoundlanders marched to Amesbury station where they entrained for Fort George near Inverness. Ten days later, the 2nd and 3rd Infantry Brigades with the Canadian Engineers moved into huts at Larkhill but at Christmas 1941 there were still 11,000 Canadians under canvas.[49] It was therefore decided to move the mounted troops and artillery into billets and in January 1915 the following areas were allotted.

> Bratton to Erlestoke: Royal Canadian Horse Artillery
> Market Lavington, Devizes and Rushall area: Divisional Artillery
> Upavon and Pewsey: Lord Strathcona's Horse
> Tilshead to Wilton: Royal Canadian Dragoons.

As a result, only the 1st Canadian Infantry Brigade remained under canvas.[50] On 4th February 1915 the Canadian Forces were reviewed by HM King George V who was accompanied by Lord Kitchener and three days later the first troops began to leave for France by way of

Avonmouth and St. Nazaire. Although there were delays and the crossing was rough and unpleasant 'there was little complaining, for present inconveniences were offset by the general feeling of relief at leaving the misery of Salisbury Plain'.[51]

In 1916 it was decided to form Australian training battalions in England from which reinforcements could be posted to Australian Divisions in France and camps for these battalions were established at Larkhill, Rollestone, Perham Down, Parkhouse and Tidworth.[52] At Rollestone, Camps 31 to 34 were taken over by No. 2 Group, comprising 4th, 5th, 6th, 7th, 12th and 13th Training Battalions which were to provide reinforcements for the 2nd and 4th Divisions. No. 5 Group consisting of the 8th, 14th and 15th Training Battalions which supplied the 5th Division, together with the Australian Artillery Depot, were stationed at Larkhill.[53] During July 1916, units of the 3rd Division began to arrive in this country from Australia and assembled at Larkhill where they remained until the Division moved to France at the end of that year.[54]

Larkhill seems to have been almost as unpopular with the Australians as the West Down camps were with the Canadians and Newfoundlanders, if the following letter represents the general opinion. Written from 'Lark Hill Camp' on notepaper headed 'Australian Commonwealth Military Forces' and dated 3rd September 1916, W.J. Sinney said.

> It has been raining like fun here and things about Lark Hill are pretty sloppy. Its a rotten place when it rains and a jolly sight worse if it keeps fine for any length of time. The dust is that fine that it will get in anywhere; do what you will, you can't get away from it.[55]

As already mentioned in Chapter 9, the New Zealand Forces were stationed at Sling and were not concerned to any great extent with Larkhill other than attending courses there or practising on the ranges. However, during the first week of January 1919, the Maoris Battalion arrived at Southampton from France and after travelling by train to Amesbury, marched to No. 5 Camp at Larkhill. The Battalion remained there until 28th February when it left for Liverpool on its way home to New Zealand.[56]

Although many different regiments were stationed at Larkhill during the First War, it remained one of the most important training centres for the Royal Artillery. In addition to instructing newly joined personnel in the class room and on the gun park, as well as on the ranges, refresher courses were provided for those who were serving in France and Flanders. However, in March 1918, life at Larkhill took an unexpected turn.

> Most of the personnel of 96 Brigade (like myself) were either wounded from France or low category men. I was with 96 Brigade until 6th March

> 1918 when we suddenly got orders to mobilise. Personnel were medically examined and most opted to be A1 for service overseas. Remounts were sent to replace and make up horse establishment. There was hustle and bustle and on 19th March we entrained at Amesbury (batteries all complete) for Southampton Docks. Then by ship to Le Havre where we disembarked and entrained for the southern sector of the 5th Army as an Army Field Brigade RFA. We were of course sent out to be in time for the great German offensive and we fought all the way back and took part in the battle of Villers – Bretonneux on 24th–25th April. The German offensive was halted there and we then went forward on 8th August with Plumer's 2nd Army till the Armistice on 11 November.[57]

For the first few months after the end of the War, a large part of Larkhill was used as a transit camp for Commonwealth troops who were awaiting repatriation. As increasing numbers returned home so the work of dismantling many of the wartime camps began, although some of the huts were retained. The first camps to be dealt with were those located to the south of the Packway (except Nos. 29 and 30), that is, Nos. 1–9 and 20–28 and also Nos. 31–34 at Rollestone. The effect on the surroundings must have been rather depressing with many buildings in various stages of being broken up, rubble and rubbish scattered over a large area and weeds taking possession of disused roadways and old parade grounds.

There was however, one encouraging sign for the future. In 1920 the School of Instruction for Royal Horse and Royal Field Artillery which had been transferred to Larkhill in 1915, was replaced by the School of Artillery.[58] At the same time the Chapperton Down Artillery School and the Heavy Artillery Gunnery School were amalgamated with Larkhill under Brig.-Gen. (afterwards Maj.-Gen.) H.W. Newcombe, CMG, DSO who had been the last commandant of Chapperton Down and who now became the first commandant of the School of Artillery. A year later the Siege Artillery School moved from Lydd to Larkhill and became part of the new establishment. The work of the School at the time of its formation has been described as follows.

> The School of Artillery, Larkhill, was responsible for providing qualified field artillery instructors for assisting in the training of Regular and Territorial units, for carrying out trials of new equipment, for stating fire-power demonstrations and for writing artillery training manuals and pamphlets.[59]

As the war years slipped away into the past, peacetime conditions returned to Larkhill but now and then its orderly routine was enlivened by some unexpected event. In the summer of 1923 a demonstration of the effect of live shells fired in 4.5in. howitzers was being given to members of the Officer Training Corps who were attending their annual camps. The target was an area of scrub and after it had been pointed out, ranging began. As soon as the second round fell, a motor cycle and sidecar was seen to leave the target at high speed much to the delight of the spectators.[60]

Although by 1920 practically the whole of the Larkhill and West Down ranges had been acquired by the War Department, restricted farming operations were permitted in certain areas. Mr. A.G. Edwards whose father was farm foreman for Mr. E.B. Maton who farmed at Enford, Upavon, Netheravon and Urchfont, was brought up at Urchfont Hill Farm (grid reference 051549) which now no longer exists. Due to its isolated position, there were few visitors and the only tradesman who called was the baker from Easterton. The children attended Urchfont School which was a mile and three-quarters distant and sometimes they would travel in a donkey and cart but otherwise they walked. When firing took place in certain parts of the ranges and it was considered that the house might be in danger, the family would be moved to a caravan that was provided for them on Redhorn Hill.

Mr. Edwards left school when he was thirteen and a half and began working on the farm: in those days his only recreation was catching rabbits. Cattle and sheep were allowed to graze on the ranges at certain times and on Urchfont Hill Farm there were three herds of cattle each of 500 head and about 5,000 sheep. Mr. Edwards who was provided with a horse, would drive one of the herds on to the ranges for grazing and when firing was due to take place in the area, he would move them to a penning on West Down. At the end of practice, he would take them back again on to the grazing area.[61]

The first steps which were taken towards the rebuilding of Larkhill after the War, were the provision of a number of married quarters for officers and soldiers and in 1928 twenty-three officers' quarters were built on the south side of Fargo Road. The area was named Strangways after Brig.-Gen. T. Fox Strangways who commanded the Royal Artillery in the Crimean War and was killed at Inkerman in 1854.

> At about half past nine o'clock, Lord Raglan and his staff were assembled on a knoll in a vain hope of getting a glimpse of the battle which was raging below them. A shell came right among the staff ... and ... blew away General Strangway's leg, so that it hung by a shred of flesh and a bit of cloth from the skin. The poor old General never moved a muscle of his face. He merely said, in a gentle voice, 'Will anyone be kind enough to lift me off my horse?' He was taken down and laid on the ground, while his life blood ebbed fast, and at last was carried to the rear.[62]

From 1929 until 1932, the appointment of Commandant of the School of Artillery was held by Brig. A.F. Brooke, DSO who subsequently commanded 2nd Army Corps of the British Expeditionary Force in 1939 and 1940, was Chief of the Imperial General Staff from 1941 to 1946 and was promoted Field Marshal in 1944. In 1929 the Commandant was also Garrison Commander and during the time that he held these appointments he put many improvements into effect with the help of the Garrison Adjutant, Capt. J.H.W.G. Richards (afterwards Brig. J.H.W.G. Richards, OBE). Huts which were no longer in use, were cleared away, road improvements were carried out and, with the help of the War Department Land Agent, Lieut.-Col. H.L. Piggot, OBE, large numbers of trees were planted, notably those on either side of the Packway.

The first church at Larkhill was a timber and corrugated iron building which was erected in the autumn of 1914 when the new camps were being laid out. This was the Church of England Garrison Church and was situated on the east side of Lawrence Road and adjoined the north-west boundary fence of the present church of St. Alban. In 1936 it was decided that a new church should be built on the north side of the Packway between Lawrence Road and Glover Road and the foundation stone was laid by Gen. Sir John Burnett-Stuart, KCB, KBE, CMG, DSO on 22nd January 1937. The Church was completed in sixteen months and was dedicated to St. Alban the Martyr by the Chaplain-General, the Rev. E.H. Thorold, CB, CBE on 21st May 1938.

Although there are no records available regarding the first Roman Catholic church, it is probable that it was built during the First World War. The Ordnance map of 1926 which was revised in 1923,[63] shows a building standing 100yds (90m.) to the north of the Packway and almost opposite what is now Lightfoot Road. This was first used as a church in about 1931[64] and is shown on the 1939 map as 'St. Anthony's Catholic Church'. This was apparently closed at some later date and was replaced by a hut on the north side of Gore Road at its junction with Alanbrooke Road. However, it was felt that other arrangements should be made and in 1965 it was agreed to build a permanent church. A site was selected on the north side of Gore Road where it crosses Wilson Road and the foundation stone was laid on 12th January 1967 by the Rt. Rev. G.W. Tickle, Roman Catholic Bishop-in-Ordinary to the Forces. The Church was subsequently dedicated to St. Barbara and St. Anthony by the Bishop on 22nd September 1968.

As in the case of many other military centres, the first social amenities at Larkhill were supplied by voluntary organizations. During the First War, the YMCA provided five huts 'in the Larkhill area' as well as three huts at Rollestone, The Bustard and Tilshead respectively.[65] On the 1926 Ordnance map a building inscribed 'YMCA Hut' is shown on the north side of the Packway about 40yds (36m.) to the east of Alanbrooke Road. In 1935 this was replaced by a brick building which was situated on the same side of the Packway but between Watson Road and Whingates Road and named in memory of Sir Arthur Yapp, KBE.[66] Since catering and recreational facilities had been included in the new barracks which were built in the 1960s, the Association felt that there was no longer any need for a YMCA centre at Larkhill.

Consequently it was decided to close this building and in 1973 it was let to the Army and became known as the Packway Hall. By the end of 1976, the building was in a dilapidated condition and the YMCA decided that it should be demolished. However, it was later agreed that the building should be given to the Army and the transfer took place in 1977. The hall was repaired and re-furnished and at the same time, was named the Sharp Hall in memory of Gen. Sir John Sharp, KCB, MC who was Commandant of the School of Artillery from 1964 to 1966. It was opened by Lady Sharp on 19th November 1978.

In the case of the Salvation Army, two huts were provided during the first War known as 'Larkhill I' which was in Larkhill Camp and 'Larkhill II' at Durrington but the exact locations have not been recorded.[67] After the War, the Salvation Army established a 'Soldiers' Home' immediately next to, and on the west side of, the YMCA on the north side of the Packway near Alanbrooke Road which has been referred to in the previous paragraph. The Ordnance maps of 1926 and 1939 both show a large building in Lightfoot Road which is inscribed 'Social Club' on these maps but it has not been possible to trace the origin of this Club.

After its formation on 1st January 1921, the Navy, Army and Air Force Institutes provided canteen facilities at Larkhill during the 1920s.[68] The Ordnance map of 1926 shows a building inscribed as 'Institute' standing between the Packway and the tennis courts which are near the present main entrance to the School and south-west of the junction of Glover Road and Bell Road. Since NAAFI canteens were usually named on large scale Ordnance maps as 'institutes', it is probable that this was the first NAAFI building at Larkhill. This is supported by fact that when the 1939 Ordnance map was published, this site was shown as an open space but the same map also marked two 'institutes', one in Whingates Road and the other in Congreve Road. Both of these were NAAFIs and are likely to have been built in place of the institute on The Packway. By 1986 the premises in Whingates Road had been taken over as a Welfare Centre but the building in Congreve Road was still in use as a NAAFI. At the beginning of 1986 the NAAFI establishments at Larkhill comprised two 'canteens', one in Shrapnel Road and the other in Congreve Road, and the Wagon Lines Club in Brackenbury Road which was opened in 1962. On the Packway, are The Packhorse Inn – a Services public house opened in February 1963 – and the NAAFI Shop which was built in 1967. Before the Wagon Lines Club opened there was another NAAFI building which was situated between Brackenbury Road and Glover Road, known as the Gun Club. When the Wagon Lines Club came into use, the Gun Club was taken over by the Administrative Wing of the School as a social centre but in 1985 it was returned to the NAAFI.

By 1915 a cinema, known as 'The Military Cinema', had been built on the north side of The Packway about 50yds to the east of Lightfoot

Road but this was burnt down in the spring of 1919 as Brig. C. Childs, CBE has recorded

> the only cinema had, a few days previously, been totally destroyed in a disastrous fire. Now only a few hundred uprights marked the spot.[69]

This was replaced by a Garrison Theatre which is shown on the 1926 Ordnance map as being on the south side of The Packway, opposite the present Garrison Church. Although this building was still shown on the Ordnance map of 1939, the word 'theatre' was omitted but another building marked 'cinema' appears on the same map on the west side of Wilson Road. This was the Globe Cinema which continued to function until it was closed in 1971 or shortly afterwards and when the necessary alterations had been made it was handed over to the Stanoc Centre in April 1974. Facilities have been made available for theatrical productions by the Larkhill Amateur Dramatic Society in the Newcombe Hall since it was built in 1966. The Larkhill Primary School which stands on the south side of The Packway between Wilson Road and Alanbrooke Road, provides for the education of young children living in the Garrison area.

One of the features of Larkhill between the wars was the Hamilton Battery which occupied a site that now lies within the perimeter fence of Purvis Lines. It was presumably named after Gen. Sir Ian Hamilton, GCB, GCMG, DSO and consisted of two 9.2in. howitzers on semi-permanent emplacements, one 8in. gun and one 6in. gun on wheeled carriages which, unlike the howitzers could be moved to a gun park when not in use. The Larkhill Military Railway passed within seventy yards of the battery position and ammunition was transferred to a narrow gauge tramway which was laid behind the guns. The quarters of the 1st Survey Company RA were not far from the battery and when firing was in progress, everyone was aware of the fact. In 1984, Mr. G.G.P. Morgan who was in the Survey Company at that time could still recall the effect.

> I often remember the booming of the two 9.2in. howitzers of Hamilton Battery situated next door, on the Balloon School side of the Survey Company. When they fired the plates and crockery used to bounce off the table.

Although a certain amount of building had taken place during the 1920s, the most active period of reconstruction before the Second World War, was from 1930 to 1939. A new mess for warrant officers and sergeants was built in 1931 while, as the leases of the old shops in The Packway came to an end, new shops built in brick were erected in their place.[70] In 1936 work began on the new Royal Artillery Mess but owing to a number of unforeseen events, one of which was the declaration of war, it was not completed until 1940 while the formal opening did not take place until 1941. This was followed in 1938 by The Packway Mess and the School Warrant Officers' and Sergeants' Mess to which extensions were added in 1964 and 1965 and new kitchens provided in 1970. The Garrison Church which has already been mentioned, was also built in 1938.

For a number of years, balloon schools were operating at Rollestone and Tilshead and further information on these will be found in Chapter12. After the First World War tented camps were erected on the area immediately to the east of Tilshead during the summer when training was carried out on West Down.

> The activities of West Down dominated the life of Tilshead in my boyhood days and, boys being boys, we spent a lot of time at 'The Camp' especially with men of the Depot Battery which was stationed at Tilshead for the whole of the summer practice season in the 1920s and early 1930s. We were given rides on horses and watched with fascination the Battery sports of tent-pegging, wrestling on horse back and musical chairs.[71]

As noted in Chapter 7, units of the Royal Tank Corps moved from Perham Down to Tilshead during the 1930s for summer training and in September 1937 the 2nd, 3rd and 5th Battalions, the 1st (Light) Battalion and the 23rd Armoured Car Company of the Royal Tank Corps were at Tilshead. Their camp was situated to the west of the track from Tilshead to Candown Farm (grid reference 044504). One September the owner of an orchard at Tilshead put a notice up at the entrance which read 'These apples are being kept for the Harvest Festival'. The next day every apple had gone and the notice had been replaced by another bearing the words 'All is safely gathered in'. Although there was no evidence as to who was responsible, it was generally considered that the operation reflected the expertise of the 3rd Battalion.[72]

The introduction of compulsory military service followed by the formation of the Militia in May 1939, brought about an immediate need for additional barrack accommodation. At Larkhill, this was provided by the construction of a new hutted camp named Newcombe Lines although sometimes known as Newcombe Barracks, which was situated on the south side of Fargo Road between the southern extensions of Willoughby and Alanbrooke Roads. After war was declared the work of the School of Artillery increased considerably as more courses were provided, new equipment such as the 25pr. gun-howitzer and the 4.5in. gun, was introduced and demonstrations and exercises were arranged. In September 1939, No.122 Officer Cadet Training Unit was established at Larkhill and this was followed, in 1943, by the 4th Royal Artillery Training Regiment (Field).

Although units of the Australian Forces were stationed at Tidworth, Bulford and Lopcombe Corner in 1940 there were none at Larkhill.

However in July of that year, sixty-eight officers and NCOs attended courses at the School of Artillery.[73] After the 2nd New Zealand Expeditionary Force arrived in this country it concentrated in the Aldershot area and although personnel attended gunnery courses at Larkhill there is no record that any New Zealand regiments were stationed there.[74] After they had disembarked at Liverpool in April 1940, the 1st Canadian Division also moved to Aldershot where they occupied permanent barracks. In October arrangements were made under which batteries moved in turn to Larkhill, for a period of ten days when, in addition to firing practice, they received instruction on the new 25prs. although these were not issued to the Division until February 1941.[75]

In the case of the Newfoundlanders, the position was rather different as authority had been given by the War Office on 22nd March 1940 for raising a Newfoundland artillery regiment – the 57th Heavy Regiment, RA – in England.[76] The first contingent of Newfoundlanders reached Liverpool on 25th April 1940 and was later supplemented by several drafts while on 16th May the name of the Regiment was changed to the 57th (Newfoundland) Heavy Regiment, RA.[77] It was stationed at Staplefield near Haywards Heath and on 15th June a second regiment – the 59th (Newfoundland) Heavy Regiment was formed.[78] In February 1941, the 57th Regiment moved to Tilshead for a week on the West Down ranges where the batteries fired 25prs. as well as 9.2in. howitzers.[79] On 6th November the Regiment was converted from heavy to field artillery and was re-named the 166th (Newfoundland) Field Regiment, RA.[80] It moved to Larkhill on 17th November where it remained until 4th May 1942.[81] The Regiment embarked for overseas on 21st January 1943.

In July 1941, the 59th Regiment moved to Larkhill for ten days for training on 12in. howitzers, two of which were issued to the Regiment and installed in gun positions at Canterbury.[82] In March 1942 the Regiment returned to West Down and on this occasion it fired 6in. guns and 9.2in. howitzers,[83] while in April, May and June 1943 it visited Larkhill and West Down when taking part in formation exercises.[84] On 5th July 1944 it landed in Normandy.

On 3rd April the first Royal visit to Larkhill took place when Their Majesties King George VI and Queen Elizabeth paid a visit to the School of Artillery and the ranges. During the War, Larkhill was subject to several air raids but the worst occurred in the early hours of 17th January 1941 when several bombs fell on Newcombe Lines. About eight huts in which men were sleeping, were destroyed, the casualties amounting to about nineteen killed and as many wounded. A number of these were in a new civilian intake which had only arrived a few days earlier. The bombs fell among huts on the west side of the camp and also on the garages, workshops and gun park.[85]

In September 1943 the 3rd Armored Division of the United States

Army moved into Warminster and several of the adjacent villages and much of their training took place on the Warminster and Imber ranges and on West Down.[86] As previously mentioned, farm tenants were allowed to graze sheep on the ranges provided that they were moved when firing was due to take place. When a flock was being driven from the ranges to a penning or fold, they would normally follow one of the many tracks which had been made by earlier generations of sheep. On one occasion when a well known shepherd – 'Truckle' Mundy – was bringing his flock back from Orcheston Down, they passed a number of American soldiers who were training in the area. The track which the sheep were following could not be seen from where the troops were standing but the shepherd knew that at a certain point, it turned sharply to the right. Just as the leading sheep reached the turn, 'Truckle' shouted at the top of his voice 'Right wheel' and the line of ewes immediately turned to the right. The Americans were amazed.

'Say, do you train your sheep to do that?' they asked.

'Of course,' answered Mundy, 'they'll do anything I tell 'em.'[87]

Although during the First World War, units of almost every arm of the British Army were stationed at Larkhill, in the Second War it was very largely occupied by the Royal Artillery or by units which provided allied technical services. There were however some exceptions and one of these was the 21st Independent Parachute Company which was formed at Larkhill in June 1942. After completing a year of extensive training on the Plain, the Company left for North Africa in April 1943.[88] As the war continued so the demand for more accommodation increased and in 1943 Fargo Hutted Camp was built. This was situated between The Packway and the continuation of Fargo Road which lies to the west of Willoughby Road and was approximately midway between the Packway Mess and the Fargo Ammunition Compound. In November 1944 it was taken over as a prisoner-of-war camp and although it was shown on a map prepared by the Commander Royal Engineers dated 25th March 1957, it was no longer marked on an Ordnance map of 1961.[89]

After the end of the War, Larkhill gradually returned to peacetime conditions and one of the first signs of this new way of life was the opening of the point-to-point course on Knighton Down when the first meeting was held on 23rd April 1947. Almost six years later HM the Queen and HRH the Duke of Edinburgh visited the course when they attended the United Services Point-to-Point on 21st February 1953.

During the late 1950s it was decided that the time had come to replace the old timber and corrugated iron huts, many of them dating from the First War, with modern buildings. A re-building programme was drawn up and on 27th September 1960, FM Viscount Alanbrooke, KG, GCB, OM, GCVO, DSO laid the foundation stone of the first building – the Soldiers' Dining Hall – on the north side of Brackenbury Road. The name of the barrack area in which the new building was located was also changed from Stirling Barracks North to Alanbrooke Barracks. This was followed by soldiers' married quarters in 1962, Roberts Barracks in 1964 and the School Headquarters, the Instructional Block and the Newcome Hall in 1966. On 6th October of that year, the new School buildings were officially opened by The Master Gunner, Gen. Sir Robert Mansergh, GCB, KBE, MC. In 1967 the Young Officers' quarters, the Warrant Officers' and Sergeants' Mess Annexe, the Womens' Royal Army Corps quarters and the medical buildings were completed. By October 1968 most of the remaining buildings including the Garrison Roman Catholic Church, had been finished.

During this spate of building activity there was one important event which was unconnected with bricks and mortar. This was the first display of guns and equipment, followed by a demonstration of artillery fire, to which the public had been admitted and was held on 27th and 28th July 1962. It was originally designated 'Larkhill Day' but in 1970 it was re-named 'Artillery Day'. The year 1970 was also the 50th Anniversay of the founding of the School of Artillery and the occasion was celebrated in numerous appropriate ways.[90] This year was also notable for the amalgamation of the original School of Anti-Aircraft Artillery at Manorbier near Pembroke which in 1961 had been re-named 'The School of Artillery, Manorbier' with the School of Artillery at Larkhill. Finally the end of the Anniversary Year was marked by HM the Queen conferring upon the School the honour of the title 'The Royal School of Artillery'.

On 3rd April 1973 the School was again honoured by HM the Queen when she paid an official visit to Larkhill. After inspecting some of the guns and other equipment and meeting members of the School Staff and their families, the Queen lunched in the RA Mess and then attended the Royal Artillery Hunt Point-to-Point. It was a day to remember.[91] HRH the Prince of Wales paid an informal visit to the School on 25th May 1979 when he watched the Battle Group Commanders' Course in progress.[92]

In January 1972 a Heavy Air Defence Compound was built on the west side of Brackenbury Road near its junction with The Packway, in order to house the Thunderbird 2 Guided Missile System. When Thunderbird was withdrawn in 1977, the enclosure became known as the Anti-Tank Compound but on 16th November it was named Purvis Lines. This was in honour of Brig. R.H. Purvis, CBE who was the last Commandant of the School of Artillery, Manorbier and who appropriately performed the naming ceremony. In 1985 these Lines were occupied by 176 (Abu Klea) Field Battery, RA.[93]

Some twenty miles north-west of Larkhill and one and a half miles south of Bradford-on-Avon, stands Midway Manor. While it is not on the Plain it is, nevertheless, of considerable interest especially to Gunners as it was the home of Lieut.-Gen. Henry Shrapnel who was

(*L. Campbell*)

42. Married soldiers' quarters in 'C' Lines which ran between Haig Road and Horne Road and were later named Arras Lines. Photographed in the First World War.

(*T. L. Fuller & Sons*)

43. The old market which was situated between Bond Street and Mons Avenue and formed part of 'Tin Town'. Photographed just before demolition in August 1969.

(*L. Campbell*)

44. A prize winning cookhouse of the First World War. The wording on the board reads: 'Our Cooks No. 1 Camp Sling. First Prize £5 – 1916'.

(*F. Stone*)

45. A 'YMCA' mobile canteen (Tea Car No. 16) of the Second World War.

(*T. L. Fuller & Sons*)

46. The camp on the west side of what was later Carter Barracks. The Kiwi can be seen in the centre background and below it and to the right a line of carriages on the camp railway. Probably 1930.

(*Mrs. N. Dibben*)

47. The Kiwi in 1919 as seen from Sling Camp which was still occupied by the New Zealanders who were waiting to return home.

(*Col. P. W. Herring*)

48. The Kiwi in 1980 after its restoration by 249 Signal Squadron Royal Signals. The apparent difference in shape is mainly due to the angle at which the photograph was taken.

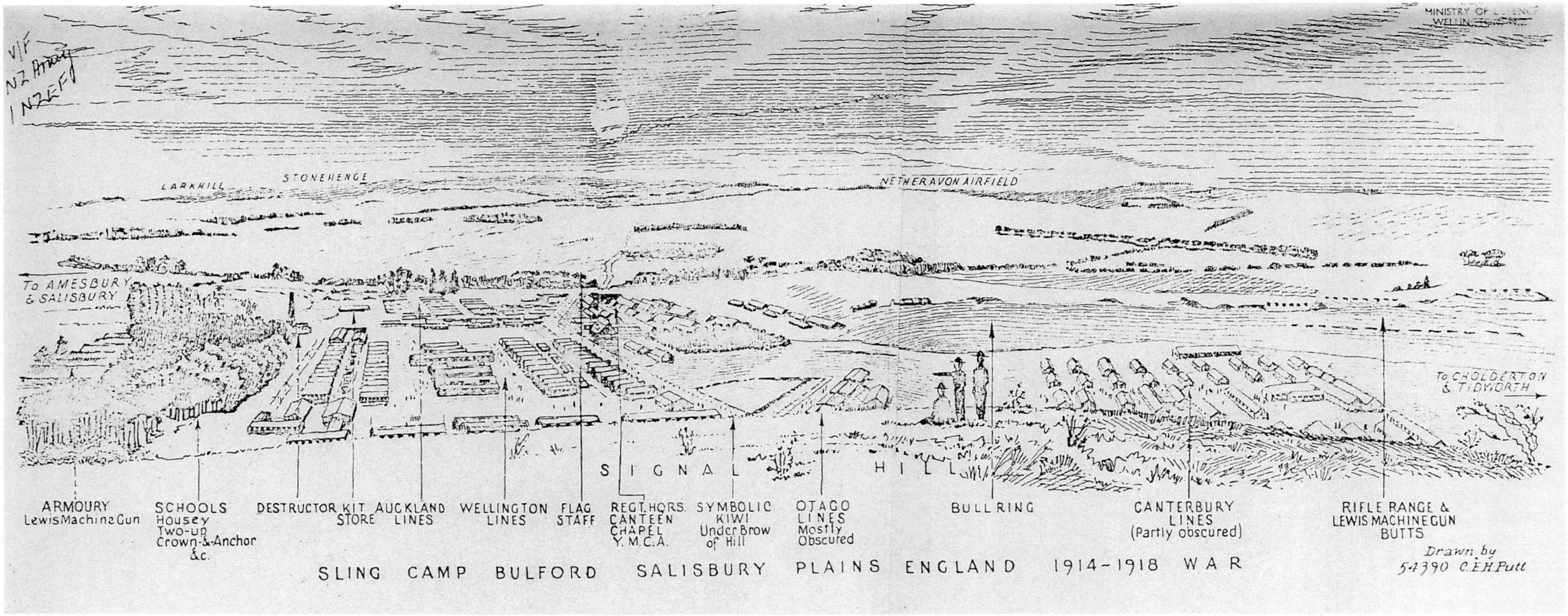

(*Ministry of Defence, New Zealand*)

49. A drawing of Sling Camp in the First World War, by C. E. H. Putt of the New Zealand Army.

T. LONG & SONS,

TILSHEAD,

PROVISION DEALERS.

ESTABLISHED 1872.

A good supply of Fresh Butter and New Laid Eggs daily.

ALL KINDS OF CAKES MADE TO ORDER.

HEAD-QUARTER STAFF, 10TH BRIGADE,
WEST DOWN NORTH,
DEVIZES, 29TH AUGUST, 1904.

Messrs. T. LONG & SONS, of Tilshead, have supplied the above Mess during their encampment at Salisbury Plain, August, 1904, with Poultry, Eggs, Butter, Cakes, etc., and must testify to the fine quality of all their goods, and the admirable way in which they have supplied me.

(Signed) FRANK PARSONS,
Steward.

(*G. H. Griffith*)

50. Testimonial advertisement for Messrs. T. Long & Sons of Tilshead in respect of their services to Headquarters 10th Brigade at West Down North Camp in 1904.

(*G. Ward*)

51. GS wagon of the Army Service Corps (now the Royal Corps of Transport) at Fargo Camp in 1913.

(*T. L. Fuller & Sons*)

52. Larkhill Camp under construction in 1915. A train on the Larkhill Military Railway can be seen beyond the partly built huts.

(*T. L. Fuller & Sons*)

53. Mud at Larkhill during the winter of 1914–15.

(*Royal School of Artillery*)

54. The book-plate of the School of Artillery (now the Royal School of Artillery) showing 'A' Mess in the centre and the School Headquarters on the right. 1926.

born there in 1761. After serving in Newfoundland he returned to England in 1784 as a lieutenant and began to experiment, at his own expense, with 'spherical projectiles fitted with bullets and bursting charges' which subsequently made his name world famous. He also compiled range tables, invented a tangent slide and made improvements in the construction of howitzers and mortars. Although he was granted a pension of £1,200 a year for life, it was considered that he did not receive the recognition that was due to him and he died on 13th March 1842 'a disappointed man'.[94]

CHAPTER ELEVEN

Warminster, Sutton Veny and Codford

As the French Revolution gathered momentum, Britain became increasingly concerned at the trend of events until finally, in February 1793, France declared war. To meet this situation, certain measures were introduced in March 1794 in order to strengthen the defences of the Country and these included the raising of 'volunteer troops of fencible cavalry ... to serve during the war in the United Kingdom'.[1] In Wiltshire, a meeting was held at Devizes on 24th April when a committee was formed to consider these proposals and a subscription list opened. Subsequently, on 4th June, it was decided to raise 'a body of yeomanry cavalry for the county' which would consist of not more than ten troops, and that the county should be divided into ten districts one of which was to be Warminster.[2] The troop was officially established at a meeting held there on 25th July when 'a sufficient number of gentlemen and yeomen having attended and agreed to join, the troop was definitely formed.'[3] This event provided what was virtually the first link between the town and the armed forces, and on 15th April 1797, the ten troops were formed into one regiment which was named 'The Regiment of Wiltshire Yeomanry Cavalry'.[4]

Early in 1798 it became known that the French were preparing for an invasion of England and immediately companies of volunteers, known as 'armed associations' were established as a precaution.[5] On 3rd July of that year 'The Warminster Loyal Association' was formed which was of a purely local character and was only concerned with 'the defence of the Town and of the country seven miles round it'. It consisted of 82 members comprising a captain, lieutenant, ensign, sergeant-major, five sergeants, five corporals, three drummers and three fifers, the remainder being privates. It would appear that they took their duties seriously since

> the Company paraded for drill every Monday and Friday at 6½ o'clock for two hours; and there was a field day every month.[6]

It is not recorded for how long the Loyal Association continued its activities but it was probably disbanded after the emergency had passed. The Yeomanry, on the other hand, continued to be very active and in January 1831 the prefix 'Royal' was conferred on the Regiment by HM King William IV in recognition of services performed during the 'Machine Riots' of 1830.[7] Some thirty years later a further honour was granted when on 23rd April 1863, HM Queen Victoria approved the additional title of 'Prince of Wales' Own'.[8]

When in 1859, relations with the French Government had again deteriorated so that another war seemed to be a distinct possibility, Volunteer Rifle Corps were formed throughout the Country. On 10th December 1859 a public meeting of 'the inhabitants of Warminster' was held when it was resolved 'that it is expedient to organize a Volunteer Rifle Corps'.[9] Between 1859 and 1860, nine rifle corps were raised in different parts of Wiltshire, namely the 1st, 2nd, 6th, 8th, 9th, 10th, 13th and 14th, the 10th being located at Warminster.[10] Its full title was the 10th Wiltshire Rifle Volunteer Corps and its officers consisted of a captain, lieutenant, ensign, assistant surgeon and honorary chaplain, the Marquess of Bath holding the rank of captain.[11] Later, the individual corps were amalgamated to become the 1st Wiltshire Volunteer Rifle Corps which after the formation of the Territorial Force in 1908, was merged with the 4th Battalion, The Wiltshire Regiment.[12]

From the time that it was established in 1859, the Corps attached much importance to rifle shooting which was carried out at Warminster.

> The site for practice was at first Knapper's Hole; the range afterwards removed to Mancombe.[13]

Knapper's Hole is shown on the Ordnance map as being situated on the north side of a re-entrant between Arnhill Down and Parsonage Farm and in 1941 the eastern end lay about 300yds (274m.) north-east of the farm buildings.[14] The name is shown on the Ordnance map of 1937 as 'Kidnapper's Hole',[15] but this is likely to be a distortion of the original name. A knapper is one who breaks or knaps stones and the original 'hole' had probably been a site for stone breaking for many years.

Since *The History of Warminster* states that rifle practice was 'at first' carried out at Knapper's Hole, that is, when the Corps was raised in 1860, and that the range was 'afterwards' moved to Mancombe, it is probable that the change of location had taken place by 1870. Two separate targets were built near the south-east corner of Mancombe Wood at grid reference 893468 but these were subsequently replaced by butts of a more orthodox pattern. The range was laid out along the line of the present entrance road to the School of Infantry and firing points are shown at intervals of 100yds (91m.) on the Ordnance map of 1887,[16] the longest being at 900yds (823m.). However, by 1901 the range had been extended to 1,000yds (914m.).[17]

In 1969, Maj. F. Myatt, MC found a number of spent Snider bullets in the area which surrounded the original targets,[18] the Snider rifle having been in service between 1866 and 1893. The plans for the new barracks were probably drawn up in 1934 since the Army Estimates for 1935–1936 provided for building sufficient accommodation for two tank battalions at Warminster. When the layout was being prepared, it was probably felt that the line of Mancombe Range formed the most suitable features on which to base the proposed development and this would account for the present disposition of the buildings. Work on the barracks began in February 1937[19] by which date the range had been closed.

After war was declared on 4th August 1914 large numbers of troops were concentrated in the Warminster area, chiefly at Codford and Sutton Veny and this is referred to later in this chapter. By 7th August the Warminster, Salisbury and Rood Ashton Troops of the Yeomanry 'were quartered in the town' and the allocation of horses took place in Portway during the same week.[20] Those who could remember Warminster during the First World War, recalled that a large Army Service Corps depot was established in Station Road but the site has not been identified. On 25th June 1915, HM King George V visited Warminster where he inspected the 26th Division the greater part of which was stationed at Codford although some units were in Warminster.[21] In August the 26th Divisional Train ASC held its regimental sports 'on a ground north of Cop Heath Lane'[22] but by the middle of September the Division had left for Salonica. It was followed by the 34th Northumbrian Division but early in 1916 this Division was sent to the Western Front where it took part in the Battle of the Somme and remained in France until the end of the War.

After the Armistice in November 1918, the military presence in Warminster was reduced once again to a troop of the Royal Wiltshire Yeomanry and a detachment of the 4th Battalion of the Wiltshire Regiment (Territorial Army). Although the Yeomanry had been in camp at Warminster in 1902[23] and it is very probable that tented camps were established there during the First World War, very little has been recorded about these. However during the 1930s, summer camps were held on the land which extends from the married officers' quarters at Elmhill, northwards across the road to the sports ground adjoining Hardinge Mess. Photographs show that these camps were occupied by units of the Royal Tank Corps but whether other regiments also made use of them is not clear. An Ordnance map published in 1937 shows the site overprinted with the words 'Warminster Camp'.[24]

In 1927 the War Office began a new programme of land acquisition which was mainly concerned with the area lying to the west of the

Shrewton–Tilshead–West Lavington road (A360). The first purchase which was completed on 29th September 1927 comprised 936 acres (379 hectares) in the parishes of West Lavington and Imber. This was followed by the acquisition of Manor Farm, Chitterne in November and Ladywell Farm, Edington and Imber Farm in December, the total area purchased during the year amounting to 3,661 acres (1,481 hectares). In 1928 Boreham Farm, Bishopstrow, Manor Farm, Upton Scudamore and further land at Imber were bought and these added another 2,435 acres while the purchase of other farms during 1929 and 1930 brought the total acquisitions from 1927 to the end of 1930 to 13,204 acres (5,346 hectares). In 1931, 1933, 1935, 1936 and 1938 an additional 13,954 acres were bought so that the area acquired between 1927 and 1938 amounted to 27,158 acres (10,996 hectares).[25]

When the acquisition of this area was in progress, a body of opinion was emerging in certain quarters of the higher command which advocated a far greater tactical use of the tank and the expansion of mechanization. One of the results of this was the formation of an Experimental Armoured Force to which reference has already been made in Chapter Six, and this probably had a bearing on the decision to develop Warminster as a centre for the Royal Tank Corps. The existence of a large area of good 'tank country' in the immediate vicinity was, no doubt, another factor that was taken into account.

An indication of the new interest that was being taken in Warminster was the inclusion in the Army Estimates for 1933–1934 of an item for building a permanent camp there. This related to Boreham Camp which was situated on the north side of Woodcock Road and later became Boreham Barracks. The Estimates for 1935–1936 however provided for a considerably larger undertaking in the construction of two new barracks and work began on these in February 1937, the contractors being Messrs. W.E. Chivers & Sons, Ltd., of Devizes. The site was divided into three sections; the headquarters with the administrative buildings and two independent barracks which were designed to provide accommodation for the headquarters of a tank brigade, two tank battalions and the supporting units. When the first barracks which stood on the eastern half of the site, were completed, they were officially opened by the Secretary of State for War, the Rt. Hon. Leslie Hore-Belisha, MP on 23rd September 1938 and named Swinton Barracks in honour of Maj.-Gen. Sir Ernest Swinton, KBE, CB. The second barracks which were built on the western portion of the site, were named Elles Barracks after Lieut.-Gen. Sir Hugh Elles, KCB, KCMG, KCVO, DSO both of these officers having been closely connected with the development of the Royal Tank Corps from its earliest days.[26]

The decision to build the new barracks at Warminster was of considerable importance to the town and the *Warminster Journal* described the opening ceremony as a 'red letter day for Warminster'.

> The occasion was an important one for the town since, as already indicated in these columns, the Barracks will accommodate two Battalions of the Royal Tank Corps, of which one – the 3rd Battalion from Lydd, Kent – has already entered into occupation, and will eventually become the Headquarters of the 1st Tank Brigade and a quota of signals, ordnance, etc.
>
> A large number of the townspeople made their way to the Barracks in order to be present at the opening ceremony and the entrance to the Barracks was gaily decorated with streamers of pennants while the flag of the 3rd Battalion fluttered proudly from the administrative block.[27]

After including a list of the principal guests and senior officers, the *Journal* reported the address of welcome by Mr. E. G. Durbin, Chairman of the Warminster Urban District Council to the Secretary of State.

> Mr. Durbin said that this day was a memorable one for Warminster, marking a stage in their history, when from being a quiet rural town they became a military centre. Warminster had experienced military occupation before, because during the Great War many thousands of troops had received their training in that area. Soon after 1918, Warminster resumed its former serenity and quietness, but now again they had come into prominence as a military town. This time however, the military occupation was of a permanent nature and they were told that these new barracks were only the commencement of a much larger scheme.[28]

In his reply, the Secretary of State emphasized that the new barracks incorporated all the latest developments in Army accommodation. They included the first example of a newly designed barracks building known as a 'Sandhurst Block' which consisted of a central section with four wings. Kitchens and dining rooms were provided on the ground floor of the central portion while the remainder of the building comprised sleeping quarters and separate sitting rooms. There were, in addition, bathrooms, lavatories, drying rooms and equipment cleaning rooms on each floor with central heating throughout the building.[29]

Work on Elles Barracks continued for some time after Swinton Barracks had been completed and the following note appeared in *The Tank* in September 1939.

> The (2nd) Battalion has just returned from leave and we found that the weather has taken a change for the better which, even though perhaps temporary, is helping us to bear more easily what we hope will be our last weeks under canvas.[30]

In 1938 work began on the construction of 23 quarters for married officers at Oxendene and 89 married soldiers' quarters in Firbank Crescent and these were completed in 1939. After war was declared, the 2nd and 3rd Battalions left Warminster, their place being taken by the 57th Training Regiment, Royal Armoured Corps. Between September 1939 and August 1940, nine of these Training Regiments were formed at the following locations.

51st, 56th	– Catterick
52nd, 58th	– Bovington
53rd, 59th	– Tidworth
54th	– Perham Down
55th	– Farnborough
57th	– Warminster

Later the 60th and 61st were formed and were stationed at Tidworth.[31]

During 1939 a Royal Army Ordnance Corps Workshop was established on a site to the north of the Great Western Railway line from Westbury to Salisbury and from which sidings were laid. This became known as the 'RAOC Depot, Warminster' and its principal task was to maintain the tanks and equipment of the 2nd and 3rd Battalions, Royal Tank Corps. Early in 1940, the RAOC Field Workshop which was stationed in the Depot, was posted to France and the remaining RAOC personnel operated the Workshop until after the evacuation from Dunkirk. When the 1st Guards Armoured Division was formed in 1941, the Workshop was re-named the 1st Guards Armoured Division Workshop and at the same time the buildings were enlarged while further developments took place in 1943. Later in the same year the workshops were transferred to the American Forces when they arrived at Warminster, but were handed back in 1945. At the end of that year, the workshop became 27 Command Workshop, Royal Electrical and Mechanical Engineers but on 1st April 1985 it was designated 27 District Workshop, REME.[32] Today the workshop undertakes general repairs to the vehicles of all three Services in South-West District as well as the complete overhaul of Centurion tanks.

On many occasions the ranges at Warminster have been used for demonstrations of fire power and one which was to be attended by the Prime Minister and senior officers of the Army and Royal Air Force, had been planned for 14th April 1942. The object was to show the effect of low level attacks by fighter aircraft equipped with cannon and machine guns, on troops and vehicles in the open. A full dress rehearsal was held on the previous day when the spectators included Lieut.-Gen. Sir Gerald Templer (afterwards Gen. Sir Gerald Templer, GCB, GCMG, KBE, DSO), representatives of the Army and the Royal Air Force and a small number of civilians. As the six aircraft of No.175 Squadron were carrying out their attack, the last pilot dropped behind and on reaching the target failed to identify it. Instead, he fired at the spectators stand, killing twenty-five and wounding seventy-one. Although the Court of Inquiry considered the pilot was to blame, there were other factors which had a direct bearing on the matter. The pilot had had little experience in firing at targets sited on the ground; he had not flown over the area before or taken part in any rehearsal; he had not been fully briefed; the spectators enclosure was not clearly marked while visibility was affected by haze. Ten weeks later the pilot was shot down over Cherbourg and reported missing.[33]

In 1943, Warminster provided the background for the production of a training film and the following is a description of this auspicious occasion by Maj. J.J. Thornton, TD.

I was with the 1st Battalion, The Highland Light Infantry in 42nd Armoured Division at Warminster during the summer of 1943. The armoured brigade was stationed at Warminster Barracks and we were in a hutted camp – now vanished – near Longbridge Deverill. A film, 'The Way Ahead' was commissioned by the War Office as a sort of semi-documentary, originally intended as a training film, so, to increase the appeal to the troops, they got Gaumont-British to make a prestige picture with all the talents of the time. Eric Ambler and Peter Ustinov wrote the script, Carol Reed directed and the players were all well-known faces including Stanley Holloway, William Hartnell, Renee Asherson, John Laurie and, not least, David Niven.

Niven had been commanding a 'phantom' detachment (Rifle Brigade) in the desert but he was whisked home to make the picture and sent down to Warminster with a film unit to shoot some background and establishing scenes. The first evening he was the guest of the Lothian and Border Horse at their Mess in the barracks when they succeeded in drinking the place dry; as a result he was not all that fit the following morning and got a very large rocket from Carol Reed. He came to our Mess that evening and, as far as I can remember, had only a couple of drinks the entire time. The power of a movie director! I suppose 1st HLI was involved because we were there and also David had been commissioned into the Regiment.

The film got the use of (the war effort being apparently temporarily suspended) one squadron of tanks, three army co-operation aircraft, one company of infantry and some guns. The shooting took two or three days in a state of continually-mounting chaos. Tanks drove hither and yon, troops marched sullenly to and fro, the sun never shone, or, if it did, did so from the wrong place – nothing seemed to be co-ordinated. In the end, it didn't really matter much for, if you see the film, you will find that all the effort was cut to about 7 seconds screen time. The Jocks got £1 a day, but the officers got £3. This was of course, unfair: the troops were allowed to keep their money, we officers were required to donate the cash to the Regimental Fund!

My lasting memory is of Niven standing in a hedge getting dressed up in all the FSMO gear, webbing and pouches, gas-mask and so forth, all supplied at enormous expense by the War Office. He says: 'The place is black with troops, there are tanks and guns everywhere, but no tin hat for Niven!' I lent him mine, but not even that appeared (least of all me) in the film.[34]

Another event in 1943, was the departure of 57th Training Regiment RAC to Catterick so that the barracks at Warminster could be handed over to the United States Army as part of the arrangement under which the Americans were to train on Salisbury Plain. In September the US 3rd Armored Division arrived in this country and moved to Warminster, Sutton Veny, Codford, Longbridge Deverill, Fonthill Bishop and Mere in Wiltshire and Cuckington, Frome and Wincanton in Somerset.[35]

The most interesting feature of the area which lies between Warminster and Tilshead is the now derelict village of Imber. Although it is recorded in the Domesday Survey of 1086 under the name of Immemerie, there had been many earlier settlements in the neighbourhood.[36] Throughout the years Imber remained a remote community on account of its situation, since it lay some distance from the traditional highways which provided a link between local towns. The old road from Westbury to Salisbury passed Bowl's Barrow a mile and a half south-west of the village while the Bath–Salisbury road lay a mile and a quarter to the north-east. Apart from this it was six and a half miles across the downs to Warminster and four to Tilshead. Sir Richard Hoare, when describing Imber in 1822 remarked that 'it is unfortunate with respect to its roads: no turnpike road passing through or near it.'[37] Its isolation was probably due to the fact that the village occupied a sheltered position at the junction of three valleys, while the surrounding land rose to a height of 125ft (38m.) or so above it. Today, when approaching from the south along the road from Heytesbury, the village is hidden from view until the last half mile or so. During the early part of the nineteenth century the population steadily increased.

> The village contains about 60 houses; its population, in 1801, was 331; in 1811, it was 379; and in 1821, 414.[38]

However, by 1931 it had fallen to 152.[39]

Isolated communities are invariably self-supporting and Imber was no exception since in the village could be found a blacksmith, carrier, carpenter and undertaker, miller, cobbler, grocer and baker. The spiritual welfare was provided by the Vicar and a Baptist preacher while the social needs were, to some extent, dispensed by the landlord of The Bell.[40] The most imposing house was Imber Court but many of the cottages were built of cob and thatched and to judge from old photographs it must have been a most attractive village. By the early 1930s the War Department had purchased the whole of the village with the exception of the property belonging to the Church Commissioners, the chapel, the school and the Bell Inn. Although the use of the ranges increased considerably after the War started in 1939, steps had already been taken to protect the village by establishing a safety area of 1,000yds (914m.) around it.

However, after the United States Forces arrived at Warminster in September 1943, it was decided to develop the area as an artillery range and this brought into question the safety of the villagers.[41] It was considered that the surest way of overcoming this problem was for those living at Imber to leave their homes and move elsewhere and this took place on 17th December 1943.[42] On 11th December a sale of sheep, cattle, heavy horses and implements took place at Seagram's Farm (originally East Farm) in Imber.[43] On 15th December a sale of 4,300 sheep was held at Edington and the auction catalogue headed 'Forced Sale of Sheep' stated that

> John Jeffrey & Son have been favoured with instructions from several farmers in this district who are selling their entire flocks of sheep (whose land has been requisitioned for Military Training). The Sale is being held at very short notice as all stock must be cleared from the land by December 17th.[44]

It has been said that when the villagers left they were told that they would be able to return after the War but by the time it had ended in 1945, the village had suffered a great deal of damage.[45] Most of this was due to shell fire and other military activities but the wind and the rain had also played their part. No doubt this damage could have been repaired or new buildings put up but this was not to be. It was not long before the War Office announced that the surrounding land would still be used for training and that it would not be possible for the villagers to return.[46] This caused a great deal of resentment and as a result the Association for the Restoration of Imber was formed which was still in existence in 1985.[47] Part of the village is now used for training troops in house to house fighting and a number of practice buildings constructed of concrete blocks have been erected. Except in a few cases such as the church and the shell of Imber Court, few of the original buildings are still standing and those which have survived are in a very derelict condition. As each year passes the old buildings crumble away a little more and the weeds establish themselves more firmly.

During the Second World War battle schools were formed in different parts of the country, the principal one being the Northern Command Battle School which was established at Barnard Castle in 1941. From 1943 it used the title 'School of Infantry' although it was not established by the War Office as an army school for infantry until 1945. During that year it moved to Warminster where it continued under the title of 'The School of Infantry'.[48] At first only the Headquarters and the Tactical Wing were located at Warminster while the other infantry training establishments which had become wings of the new School remained where they were. The oldest of these was the Small Arms School which had been founded in 1853 as the School of Musketry at Hythe in Kent. After several changes in its title and constitution, it became the Small Arms Wing in 1950 and when Hythe closed in 1969 it was transferred to Warminster.[49] In September 1922 the Cavalry School which was stationed at Netheravon, took up new quarters at Weedon in Northamptonshire[50] and a machine gun school that had previously been at Seaford in Sussex, took its place. During the next thirty years or so, numerous changes occured at Netheravon both in establishments and designations but in 1948 the Support Weapons Wing was formed and was still at Netheravon in 1986. In view of the developments which were taking place in the field of communications, a Signal Wing was formed at Hythe in 1950 but was transferred to Warminster in 1969 while in 1973 a NCOs Tactical Wing was established at Dering Lines, Brecon in Powys.[51] The School establishment also includes the Warminster Support Unit which until 1973 was known as the Administrative Wing. This is the administrative unit for officers and soldiers who are based at the School and also for the Staff and Students of the Junior Division of the Staff College.

A Demonstration Battalion is also attached to the School which until 1978, was provided by a regular infantry battalion. However, from that year until 1982, an Infantry Demonstration Battalion (IDB) was formed as a composite unit which was made up of personnel drawn from different battalions. In 1982 the IDB was disbanded and its role re-assumed by a Demonstration Battalion which in 1986 was The Worcestershire and Sherwood Foresters Regiment (29th/45th Foot). For a number of years the Demonstration Battalion was quartered at Knook Camp which lies about a mile to the east of Heytesbury but when Boreham Barracks (later re-named Battlesbury Barracks) were rebuilt in 1964 they were allocated to the Demonstration Battalion which then moved from Knook.[52]

An armoured squadron, detached from its parent regiment in Tidworth is also stationed at the School and in 1986 this was provided by the 5th Royal Inniskilling Dragoon Guards. These squadrons are located in Harman Lines which were built above Sack Hill in 1978 and are named after Gen. Sir Jack Harman, KCB, OBE, MC who was Colonel Commandant of the Royal Armoured Corps at the time. Artillery support is provided by a battery of 14th Field Regiment RA stationed at Larkhill. To summarize the position as it was in 1986, the School of Infantry comprised Headquarters, Tactics Wing, Signals Wing and Support Unit at Warminster, with the Support Weapons Wing at Netheravon and the NCOs Tactical Wing at Brecon.

A welcome sign of the return to peacetime conditions, was the formation of the School of Infantry Beagles in the spring of 1948 by Col. (afterwards Brigadier) W.C.P. Richardson, DSO who was Commandant of the Tactical Wing. Maj. (later Lieutenant-Colonel) D.B. Rooke, MC an Instructor in the Company Commanders Division, became the first Master and huntsman and drafts were obtained from a number of well-known packs, including the Trinity Foot. Although the country includes the area between Warminster and Netheravon, hunting on the ranges has had to be restricted to the boundary areas in recent years. However on the other side of Warminster, the country extends to Wincanton in the west and as far as Lacock in the north.

When packs are run by the Services, Masters tend to come and go at frequent intervals but at Warminster there have been two notable exceptions. Lieut.-Col. J.S. McLaren, OBE who was Master and huntsman from 1953 to 1957, became Master again when he returned to the School in 1968 and was still a Joint-Master in 1986. Col. A.F. Mackain-Bremner, OBE was Joint-Master for the seasons 1962–1964, 1966–1968 and 1972–1975 and these two extended masterships enabled a continuing policy to be adopted with regard to hound breeding.

Although it is probable that support from service members reached its peak between 1954 and 1957, it was decided in 1955 to form a hunt committee which included civilian members. This step would appear to have been justified by the fact that since 1955 support by members of the Forces has decreased, and consequently in 1985, it was decided to change the name of the Hunt to The Wiltshire and Infantry Beagles.[53]

The establishment of the School Beagles was followed in 1951, with the founding of the School of Infantry Saddle Club by the Commandant, Brig. R.E. Goodwin, DSO (afterwards Lieut.-Gen. Sir Richard Goodwin, KCB, CBE, DSO), Lieut.-Col. J. Mogg, DSO Chief Instructor (afterwards Gen. Sir John Mogg, GCB, CBE, DSO) and Maj. J.B. Dye, MC, Officer Commanding the Platoon Commander's Division (afterwards Maj.-Gen. J.B. Dye, CBE, MC). To begin with, one horse was bought and a lorry converted into a horse-box so that members could have a day's hunting while later, another horse and some ponies were acquired. When Brig. F.E. Kitson, OBE, MC (afterwards Gen. Sir Frank Kitson, KCB, CBE, MC) became Commandant in 1973, the Club began to expand and in 1976 the name was changed to The Infantry Saddle Club. This was due to the fact that the Club was now running courses for Army personnel who wished to learn to ride, irrespective of where they were stationed. Since then the Club has continued to flourish and in 1986 it had twenty horses and ponies, indoor and outdoor riding schools, as well as show jumping and cross country courses. It was also running some twenty courses a year, each of a week's duration. Two classes of membership are available: Full Members who are stationed in the Garrison and their families and Associate Members who are serving but are not in the Garrison and civilian members employed by the Ministry of Defence and their families.[54]

The 1950s saw the beginning of the post-war building activities in the Garrison area and forty married officers' quarters were completed at Elmhill in 1953. These were followed by fifty married soldiers' quarters in Houldsworth Avenue and sixty-two in Wilson Square, all of which had been finished by the end of 1955.[55] Boreham Hutted Camp adjoining Woodcock Road, was replaced by permanent buildings and re-named Boreham Barracks in 1957 but, as already mentioned, these were enlarged in 1964 and designated Battlesbury Barracks. However it was between 1965 and 1969 that the principal building developments took place, in order to provide accommodation which had become necessary for a number of reasons. In 1943 a Director of Infantry had been appointed with headquarters at the War Office but in 1965 the Directorate of Infantry moved from London to Warminster.[56] At the same time the Infantry Trials and Development Unit moved to the School but these changes were only a beginning. When Hythe finally closed in 1969, the Small Arms Wing and Signals Wing were transferred to Warminster[57] while during the same year, a Junior Division of the

Staff College was formed at the School in order to train junior officers of all arms in tactics and fundamental staff duties.[58] Thus the urgent need for new buildings and accommodation during this period can be readily appreciated.

Not only were a number of new buildings erected but several which had been built between 1937 and 1940 were altered or put to a different use. The original sergeants mess in Swinton Barracks is now occupied by the Infantry Trials and Development Unit while the battalion headquarters in Elles Barracks has become the Weapons Museum. As well as providing new administrative buildings, work began in 1965 on the construction of additional quarters for married soldiers in Pepper Place and continued at intervals until those in Goodwin Close were finished in 1974, a total of 188 having been built. The roads which serve the married quarters are named after the following officers who were Commandants of the School during the years which are shown after their names.

Wilson Square	Maj.-Gen. T.N.F. Wilson, CB, DSO, MC	1942–1943
Houldsworth Avenue	Brig. Sir Henry Houldsworth, KBE, DSO, MC, TD	1943–1944
Pepper Place	Brig. E.C. Pepper, CMG, CBE, DSO	1944–1945
Firbank Crescent	Maj.-Gen. C.L. Firbank, CB, CBE, DSO	1948–1951
Goodwin Close	Lieut.-Gen. Sir Richard Goodwin, KCB, CBE, DSO	1951–1954

In the case of houses for married officers, fourteen were built at Oxendene in 1966, immediately to the north of those which had been completed in 1939 while another four were added to Elm Hill adjoining the eastern boundary of the 1953 development.[59]

Although there had been a small temporary church in the Garrison area it was not until 1966 that the question of building a permanent Garrison Church received serious consideration. Subsequently on 29th January 1967, the foundation stone of the new church was laid by Lieut.-Gen. Sir Anthony Read and the occasion has been recorded on a stone, set in the wall near the north entrance, which bears the following inscription.

To the Glory of God

This foundation plaque for St. Giles' Church
was unveiled on 29 January 1967 by
Lieutenant-General Sir Anthony Read, KCB, CBE, DSO, MC
Commandant of the School of Infantry 1959–1962

This stone was formerly part of the Parish Church
of the same name, in the village of Imber

Commandant: Brigadier A.J.A. Arengo-Jones, OBE

Chaplain: The Reverend Geoffrey Higgins, MA, CF

The Church was completed in the following year and was dedicated to St. Giles on Trinity Sunday, 9th June 1968, by the Venerable Archdeacon J.R. Youens, CB, OBE, MC, QHC.[60]

When Hythe closed in 1969, the exceptionally fine collection of weapons which the School of Musketry and its successors had assembled since 1853, was transferred to Warminster. Originally the exhibits comprised, for the most part, muskets and rifles but in recent years the scope of the collection has been expanded so that it now includes pistols, automatic weapons and anti-tank weapons. The exhibits are displayed in the Weapons Museum which was officially opened on 16th May 1970 by General Sir Anthony Read and can be visited on weekdays by appointment. Building work continued into the 1970s and while the soldiers' married quarters in Goodwin Close were being completed in 1974, another eleven married officers' quarters were added to Elmhill. In 1978, Harman Lines which have already been mentioned, were built at the top of Sack Hill.

However it is the ranges which have probably created the greatest impression on those who have served at Warminster or have attended courses there. The great expanse of grass, stretching away to the horizon and only interrupted by the occasional belt of trees, is something which is not easily forgotten. The following valedictory verses were written by an anonymous National Serviceman in the Wiltshire Regiment when it was Demonstration Battalion in the early 1950s, on his return to civilian life.

It's goodbye now to the skeletal trees,
to Strip Wood's ambiguities;
For nevermore shall I embus
For Imber Clump where no birds mourn
For Wadman's Coppice left forlorn,
Or Tinker's Firs without a cuss.

And farewell, falling walls of Imber,
Your blinded windows hold no hate
But sometimes with the dead remember
A peaceful and fireside mirth
Of men who tilled a saner earth –
I leave you twice desolate.

Bye-bye bare Bottoms that I knew,
Oxendean and Breakheart! Pug's Hole
Most fascinating spot, adieu!
A veteran of Salisbury Plain,
Of many a bogus Alamein,
Out of the Army and on the dole,
I'll think of you, I'll think of you.[61]

The village of Sutton Veny lies about two miles south-east of Warminster and in 1914 when practically all Army transport was horse-drawn, its proximity to the town made it a convenient area in which to station troops. By November the work of erecting huts had begun and it was reported that 'Sutton Common was now a scene of great activity'.[62] The camps covered a large area; two were on the south side of Deverill Road adjoining The Beeches and Long Iver with three more between that road and Five Ash Lane. Another three camps were built on the west side of the Bishopstrow road and between Five Ash Lane and Marsh Lane which was known as 'The Common', while on the east side of the road there were five camps near Cooper's Bottom.[63] Other camps were built in the Sandhill area between Southleigh Wood and Longbridge Deverill.

Large quantities of timber and other materials were needed for the construction of these camps and after they had been built, stores, rations and equipment had to be supplied. In order to deal with this transport problem, the Sutton Veny Camp Railway was built from Heytesbury station to Sandhill near Longbridge Deverill and information about this line will be found in Chapter13.

The erection of the huts continued throughout the spring of 1915 and in April part of the 26th Division moved to Sutton Veny,[64] the first troops arriving on 28th April although some of the camps had not been finished.[65] Prior to this, a camp near Cooper's Bottom had been used for holding internees but with the arrival of the Division, they were moved elsewhere and the camp used for prisoners of war.[66] During the summer of 1915 the troops underwent a period of intensive training:

> Each day the different regiments would march off headed by their band and complete with field kitchens drawn by mules. The kitchens were mounted on wheels and were fitted with a fire grate and chimney so that the food was cooked as the column marched along.[67]

The fields on the south side of the road (B3095) near Norton Waters were used for trench digging exercises:

> The whole area was a mass of trenches which were used to practice the conditions of trench warfare. The troops would live in these for a week or so and all food was brought up in dixies as it would have been, in action.[68]

The principal forms of relaxation available to the troops, consisted of the YMCA which was established in Sutton Veny House, then known as Greenhill House, and three cinemas. One of these was in Dymocks Lane and the other two which were Army cinemas were in Five Ash Lane and in a quarry near the junction of the Lane with Deverill Road.

> The quarry one came to an untimely end. It was managed by a couple. Mrs. was fourteen stone plus, Mr. eleven stone minus. One evening the projector broke down and Mrs. ordered Mr. to take the money and disappear. Being an obedient husband he did so and she told the troops she could not find him, so she was unable to return their money but would give them

a free ticket. This was no good as some of the lads were off to France next day. When she persisted, they set fire to the place and a young subaltern tried to restore order. He received a couple of bricks, so wisely faded away. The troops dragged the piano outside and sang songs ending with 'Keep the Home Fires Burning'. They then threw the piano on the fire and departed.[69]

By the end of July 1915 a church had been built in the camp for members of the Church of England and a chapel for those who were Nonconformists was in course of construction. In the middle of September, the 26th Division proceeded overseas and, no doubt, life became less demanding for those who remained as the officers of the RAMC who were stationed at Sutton Veny were lent the Wick and District Beagles for the season.[70] One of the hazards of the Sutton Veny camps was the risk of runaway trucks on the camp railway and a report in the *Warminster Journal*[71] referred to 'trucks standing stationary at the top of the incline on the military light railway' which suggests that they did not start of their own accord. Once loose trucks started to move, there was nothing to stop them until they reached the main line at Heytesbury Station but even then they sometimes went on down the main line. The fact that the report was headed 'More Runaway Trucks' suggests that it was not an unusual occurrence.

The hospital which opened in 1916, was officially described as a 'hutted hospital' and contained beds for 11 officers and 1,261 soldiers.[72] In December 1916, No.1 Australian Command Depot which has been referred to in Chapter 6, moved to Sutton Veny where it remained until October 1919, while the Overseas Training Brigade of the AIF was in camp at Longbridge Deverill.[73] This added to the work of the hospital and after the Armistice in November 1918, No.1 Australian General Hospital was transferred to Sutton Veny.[74] Although deaths occurred in the hospital from time to time, they reached an unprecedented number in 1918–1919 when a particularly virulent form of influenza which was known as Spanish influenza struck the camp. This caused a large number of deaths amongst the Australian troops and the prisoners of war, and the memorial in the Australian chapel in St. John's Church records the names of 142 Australian soldiers and 3 nursing sisters although these were not all casualties from influenza. Part of the churchyard has been formed into an Australian War Cemetery and a memorial service for those who died is still held on the Sunday nearest to Anzac Day (25th April). The Australian Liaison officer at the School of Infantry always attends and is frequently accompanied by a representative of the Australian High Commission. The Church and cemetery have many Australian visitors throughout the year.[75]

After the War, the village gradually returned to peacetime conditions, the huts were removed and, in 1919, the camp railway was lifted.[76] When the Second World War began, two camps were established at Sutton Veny, the larger one astride the Deverill road between The Beeches and Long Iver and the smaller in Best's Lane.[77] After the United States Forces arrived in this country in September 1943, the 36th Armored Infantry Regiment took over Long Iver Camp and remained there until it left for France where it landed on 25th June 1944. The 83rd Armored Reconnaisance Battalion moved into Longbridge Deverill Camp which it did not leave until the end of June, reaching France on 4th July.[78] A tented camp at Bishopstrow was also occupied by the American Forces.[79]

Some seven miles south-east of Warminster and standing on either side of the Wylye, are the villages of Codford St. Peter, Codford St. Mary, Stockton, Sherrington, Boyton, Upton Lovell and Corton each of which was affected by the Army during one or both of the World Wars. This began in September 1914 when *The Warminster Journal* announced that

> A portion of the new Kitchener's Army is coming to this district for training and during to-day and to-morrow, no less than 24,000 will arrive in the neighbourhood of Codford where they will be encamped for this purpose.[80]

A week later it was reported that 'Codford is rapidly being transferred from a village of little more than five hundred people to a military town of many thousands' and that 'camping grounds' for the tents were located at 'the Fisherton Delamere end of Codford'. These were mainly on Manor Farm although the 2nd Wessex Royal Engineers were in camp near Codford Station.[81] On 25th September *The Journal* reported that

> The camps are divided into two divisions, the 25th and the 26th. The former is situated in the Codford and Fisherton Delamere area ... while the latter is around Sherrington.[82]

A Red Cross Hospital was established near Codford St. Mary Church but by the middle of October 'erection of huts to replace tents has begun, starting on the Fisherton side'.[83]

There were fifteen camps at Codford, Camps 1 and 2 being on the north side of the Warminster road between The Manor at Fisherton and Malmpit Hill and Camps 3 and 4 on the south side of the road, opposite Foxhole Bottom. Camp 5 lay between Little Wood and Codford St. Mary Church with numbers 6 and 7 to the north and west of the Church, respectively. Camps 8 to 12 were located on the east side of, and close to, the Chitterne road while numbers 13 and 14 stood besides the road to Manor Farm, on its east side. Number 15 lay to the north of the Warminster road opposite Ashton Gifford, all these camps being occupied by the 25th Division. There were a further twelve camps spread along the Wyley Valley close to the villages of Sherrington, Boyton, Upton Lovell and Corton which were allocated to the 26th Division. Codford Military Hospital was situated in Punch Bowl Bottom and extended down to the Chitterne road.[84] In October 1914 it was announced that an area of fifty acres at Knook, some 2½ miles north-east of Codford St. Peter, had been selected for a camp[85] and according to the late Rev. E.D. Ginever, Knook was laid out as an artillery camp in 1914 while at the same time Heytesbury House was taken over for officers' quarters.[86]

As in the case of Sutton Veny, the expansion of Codford as a military centre was, to a large extent, dependent on a railway connection and a description of the Codford Camp Railway is included in Chapter 13. The sudden arrival of a large number of soldiers in a rural area was bound to have a considerable impact on village life. At Boyton 'the Church has been given up entirely to the use of the troops'[87] while at Codford temporary shops, cafés, barber's saloons and so on, sprang up on any vacant site and even in front gardens beside the main street.[88] At first the welfare of so many soldiers was a problem but by early October the situation had improved.

> The Bristol YMCA have erected a commodious marquee in the centre of Codford where something like a thousand men can be accommodated.[89]

Some six months later a YMCA hut had been built thanks to the generosity of 'an anonymous Bath donor'.

By the middle of October 1914 the fine autumn had come to an abrupt end and a long period of rain followed. The plight of the Canadians who were under canvas on West Down has been described in the last chapter but Codford also had its share of wet and cold. Extensive flooding occurred in the village especially in the area around the railway station and the camp soon became known as 'Codford-on-the-Mud'.[90] The conditions are reflected in the cartoon on page 63 which displays a rather macabre sense of humour. The 26th Division left Codford in September 1915 for Salonica and later the 25th Division went to France where it remained for the rest of the War.

Although these Divisions were replaced by other British troops, in July 1916 Codford Camp was chosen as the New Zealand Command Depot for men who, having been wounded or sick, had completed their convalescent period but were not yet fit for active service.[91] Since the Depot held some 2,500 men it was essential to provide a hospital in the immediate vicinity and Codford Camp Hospital was taken over by No.3 New Zealand General Hospital.[92] It contained beds for 10 officers and 980 soldiers.[93]

> Codford Camp was not popular for its surroundings or its exterior attractions. Less than a mile from the camp headquarters lay the little old-fashioned Wiltshire village from which the camp derived its name – the only oasis of interest in the midst of a drab environment. Men usually went to Codford after fourteen days convalescent leave which did not enhance its popularity.[94]

Under such circumstances it was particularly important to provide adequate facilities for relaxation during off-duty hours.

> Probably the most popular of Codford's institutes was the New Zealand War Contingent's Club, 'Aotea Roa' ... claimed inside and outside the

> Depot as the best equipped club on the Plain. Its management was in the hands of ten capable ladies of the War Contingent Association whose arduous work ... was very faithfully done.
>
> The ubiquitous YMCA needed no trumpeter to proclaim its good work. It had two huts in the Depot, both well equipped and popular.[95]

Aotea Roa is the Maori name for New Zealand. On 31st October 1918 the strength of the Depot was 33 officers and 2,382 soldiers while the Hospital was 49 officers and 979 soldiers making a total of 3,361 for all ranks.[96]

Little has been recorded about Australian troops at Codford but since there are the graves of 31 Australian soldiers as well as those of 66 New Zealanders in Codford St. Mary New Churchyard,[97] it would appear that some Australians were admitted to the New Zealand Hospital. The Rev. E.D. Ginever has recorded that the Australian Reserve Brigade which was stationed at Knook, placed its flags in Heytesbury Church and that members of the Brigade presented a lectern bible to the Church, when the Brigade left.[98] During the First War a reproduction of the badge of the Australian Forces was cut on the side of the hill above Foxhole Bottom at grid reference 984392 or thereabouts. Although No.1 Australian Command Depot was at Sutton Veny, this site may have been chosen on account of its prominent position. Early in the Second War the badge was covered with turf as it was thought that enemy aircraft might be using it as a guide on their way to bomb Avonmouth Docks. After the War Mr. Michael Stratton of Stockton voluntarily undertook the maintenance of the badge until 1969, when the Marquess of Bath kindly made arrangements for its restoration.[99]

After war was declared in 1939, the Army returned to Codford and a limited number of camps were established although these were mostly on different sites to those of the First War. The principal camp was on the east side of the road from Codford St. Peter to Manor Farm, opposite the sites of Camps 13 and 14 of the previous war. There was also a camp between Codford St. Mary Church and Little Wood near the site of Camp 5. In September 1941 the 6th Guards Armoured Brigade arrived at Codford and remained there until 1943. It was said that the members of the Brigade were 'probably the most popular troops to have been in the village in either war'.[100]

With the arrival of the American Forces in September 1943, the 32nd Armored Regiment was stationed in Codford St. Peter, the Maintenance Battalion and Supplies Battalion in Codford St. Mary with the 45th Armored Medical Battalion, Combat Command 'A' and Trains Headquarters in Stockton House.[101] These units left for France on 2nd–4th July 1944 and were followed in Codford by other American troops. After the war a Polish Corps under Gen. Anders occupied the various camps with Corps Headquarters in Stockton House. The Polish troops remained in Codford until they were demobilized but some men stayed in the village after leaving the Army and several were still there in 1984.[102]

CHAPTER TWELVE

Wings Over The Plain

After the end of the Franco-Prussian War in 1871, the War Department decided that more attention should be paid to the use of balloons for military purposes. The matter was accordingly referred to the Royal Engineers' Committee which had been re-organized in 1870 with a view to examining any new developments that might have a military value. However, as there was already a small committee which was concerned with balloons, it was reappointed as the Balloon Sub-committee of the main body.[1] In 1878, authority was given for balloons to be manufactured on an experimental basis at Woolwich[2] and two years later instruction in their use began at Aldershot under Captain H. Elsdale RE and Captain J.L.B. Templer, KRRC. At the end of 1882, training was transferred to the School of Military Engineering at Chatham where there was already a balloon establishment and this later became known as the School of Ballooning.[3] Progress was such that in 1885, balloon detachments took part in the Bechuanaland and Sudan expeditions while in 1889 a balloon detachment attended the Aldershot manoeuvres.[4]

The following year saw the formation of the Balloon Section and Depot at Chatham and so for the first time, a unit which was concerned with operating in the air formed a part of the British Army. In 1891 the Section was transferred to Aldershot and was joined the next year by the School of Ballooning which in 1897 was officially named the Balloon Factory, Col. Templer having been appointed Superintendent.[5] Three Balloon Sections served in the Boer War where they proved their value for long distance observation and after the War, courses covering the use of balloons were held on Salisbury Plain for officers of all arms, especially Gunners.[6] The value of balloons was beginning to be appreciated and in 1903 Lieut.-Col. J.E. Capper, CB, RE (afterwards Maj.-Gen. Sir John Capper, CB) was posted to command the Balloon Section. The appointment of this very able officer ensured that no opportunity would be lost for research and development.[7] In 1905 the title of 'Balloon Section' was changed to 'Balloon Company', and the whole Command designated the 'Balloon School and Companies' while in April 1906, the School was given official status.[8] During the winter of 1905–1906 the Balloon Factory moved from Aldershot to more spacious surroundings on Farnborough Common.[9]

However there were signs that the days of the balloon were numbered since by 1905 Lieut. J.W. Dunne of the Wiltshire Regiment had produced a number of model gliders. On seeing these, Col. Capper realized the possibilities of this new idea and arranged for Dunne to be attached to the Balloon School in order to direct the building of a prototype.[10] Tests were carried out in Scotland during the summer of 1907 when an engine and propeller were fitted to the glider and these experiments were continued in the following year.[11] At the same time an aeroplane designed by S.F. Cody was built and tested at Farnborough and on 16th October 1908 this machine flew for a distance of about 463yds (424m.) thus making the first substantial flight by a power driven aeroplane in Great Britain. Cody who was an American was the leading authority on man-lifting kites which he successfully designed and which were adopted by the Balloon Companies in 1906. He became a British subject in 1909 but was killed in August 1913 when flying one of his machines. Unfortunately these attempts to produce a satisfactory heavier-than-air machine received a set-back in April 1909 when the War Office announced that owing to the heavy costs of the trials of the Dunne and Cody machines, no further experiments with aeroplanes would be permitted.[12]

In the meantime a new balloon had been designed to replace the original spherical pattern and in May 1909 the *Warminster Journal* published a report on 'the first trials of the new Army dirigible balloon which was brought out of the Balloon Factory on Farnborough Common for its lifting and balancing tests'. The report added that 'the shape of the balloon is much like that of a whale with a three-fluked tail'.[13] The importance which was attached to the use of observation balloons can be judged from the steps which were taken to deal with those belonging to an enemy. The *Annual Reports of the School of Gunnery* for 1902, 1907 and 1909 contain accounts of the engagement of balloons with artillery fire[14] while the *Journal* for 8th May 1909 included the following report.

> Experiments in shell firing at balloons will begin early this month on Salisbury Plain. ... The 12th Brigade RFA and No. 2 Balloon Company RE proceeded to Salisbury Plain from Aldershot, on 28th inst. in connection with the firing. The Balloon Company will go into camp at Durrington while the howitzer batteries will join the practice camp at Larkhill ... The balloons to be fired at will be held captive by steel wires attached to a winding drum which forms part of the regular equipment of Balloon

Companies. Shrapnel will be used. The firing will take place within the artillery danger zone.[15]

In June 1909 H.C. Barber, more usually known as Horatio Barber, who was one of the earliest flying enthusiasts, rented a small plot of land at Larkhill, a little to the west of the junction of Tombs Road with the Packway. This apparently unimportant transaction proved to be the first link in the chain of events which indirectly led to the formation of the Royal Flying Corps – later to become the Royal Air Force. On this site he arranged for a shed to be built in which he housed his new aeroplane when it arrived in June 1909. By the end of that year he had been joined by G.B. Cockburn and Capt. J.D.B. Fulton, RFA who stabled their machines in adjoining sheds. Capt. Fulton who was serving in the 8th (Howitzer) Brigade, RFA which was stationed at Bulford, owned a Bleriot monoplane and he was soon recognized as one of the leading pilots of his time. He transferred to the Air Battalion, RE in 1911 and in 1913 became an Instructor at the Central Flying School, Upavon. By June 1915 he had been appointed Chief Inspector of the Aeronautical Inspection Department but died at an early age in November of that year. G.B. Cockburn was another civilian who was dedicated to flying and has been described as 'a philanthropic private aviator' who at his own expense taught the first four Naval pilots to fly.[16]

The War Office was now attaching greater importance to the use of aeroplanes and in May 1909 two buildings were erected for the use of the Army and the Navy, to the south of the three which have been mentioned in the previous paragraph. In June 1910 a three bay hangar was built by the British and Colonial Aeroplane Company to house the aircraft used by the Flying School which the Company had been allowed to set up.[17] These can still be seen in Wood Road opposite Colquhoun Road and in 1986 they were being used by the Royal Army Ordnance Corps as an Accommodation Store. They are probably the oldest aeronautical buildings still in existence in Britain. The British and Colonial Aeroplane Company later became the Bristol Aeroplane Company and ultimately a part of The British Aircraft Corporation. In March 1911 two additional sheds were built for The Bristol Flying School while in the following month, four more were constructed for the War Department.[18] An Ordnance map which marks the property on Salisbury Plain belonging to the War Department, shows that in 1911 there were eight buildings on the west side of Tombs Road which are inscribed 'Aeroplane sheds' on the map – Wood Road had not yet been built.[19]

The airfield extended westwards from the hangars towards the site of the old Down Barn destructor which stood at the south end of Bingham Road. A plaque recording the formation of the aerodrome has been fixed to a stone pedestal which stands on the east side of Wood

Road, about 165yds (150m.) north of Fargo Road. The inscription is as follows.

ON THIS SITE THE FIRST AERODROME
FOR THE ARMY WAS FOUNDED IN 1910 BY
CAPTAIN J.D.B. FULTON R.F.A. AND MR. G.B. COCKBURN
THIS LATER BECAME 2 COY AIR BN R.E.
THE BRITISH AND COLONIAL AEROPLANE COMPANY
FORERUNNERS OF THE BRISTOL AEROPLANE COMPANY
ESTABLISHED THEIR FLYING SCHOOL HERE IN 1910
THE FIRST MILITARY AIR TRIALS
WERE HELD HERE IN 1912.

In October 1910 the War Office announced that it had decided to expand the air branch of the Army by reorganizing the Balloon School so that it would include aeroplanes as well as airships, balloons and kites.[20] In the same month Horatio Barber moved from Larkhill to Hendon and his aircraft shed was taken over by the Flying School. His departure did not alter the fact that as regards flying, Larkhill was developing into a place of some importance as the *Salisbury and Winchester Journal* reported.

> Salisbury Plain as an aviation centre is increasing in renown and the next few years may see great developments in the art, in this district ... A long line of hangars or sheds has been erected on Larkhill wherein are housed a variety of flying machines including the biplanes of the British and Colonial Aeroplane Company.[21]

In accordance with an Army Order issued on 28th February 1911, the Balloon School was re-formed as the Air Battalion Royal Engineers with effect from 1st April of that year.[22] It consisted of Battalion Headquarters, stationed at Farnborough; No. 1 Company which was responsible for airships, balloons and kites and was also at Farnborough, and No. 2 Company which was concerned with aeroplanes and was located at Larkhill[23] under the command of Capt. Fulton. At the same time the Balloon Factory was re-named the Aircraft Factory.

Matters were now proceeding apace and the Committee of Imperial Defence was asked to consider the whole question of aviation in the services, with a view to the formation of an air force. On 18th December 1911 a subcommittee was appointed to look into the matter and this in turn delegated some of its work to a technical subcommittee. In spite of this extended chain of responsibility, on 25th April 1912 the Imperial Defence Committee approved the recommendation which these two subcommittees had put forward. Consequently on 13th May 1912, the Royal Flying Corps was formed and it was also directed that this new arm should include a military wing, a naval wing, a central flying school, a reserve and the Aircraft Factory at Farnborough.[24] One of the results of these changes was that the Air Battalion no longer existed and No. 1 Company (airships and balloons) became No. 1 Squadron,

RFC which remained at Farnborough while No. 2 Company (aeroplanes) became No. 3 Squadron RFC and continued at Larkhill. At the same time steps were taken to form No. 2 Squadron from the reserve of pilots at Farnborough.[25] No. 3 Squadron was therefore the first Squadron to fly aeroplanes and it is still in very active existence as No. 3 (Fighter) Squadron, RAF. During the Falklands War it flew 207 operational sorties and, as the oldest squadron in the Royal Air Force, bears the distinguished motto *Tertius Primus Erit* – The Third shall be First.

Action on these various decisions was not long delayed and the Central Flying School was officially opened on 19th June 1912. With the exception of eleven hangars all the buildings were located on the north side of the Upavon–Ludgershall road (A342) and a little over a mile to the east of the village of Upavon. Eleven hangars were built on the south side of the road adjoining the main airfield with a further six along the eastern perimeter on the north side of the Ludgershall road. The choice of the site was subject to some criticism in view of its exposed position, the buildings being located at between 550ft. (168m.) and 625ft. (190m.) above sea level. It was also felt that the main airfield left much to be desired, since it sloped to the south-west from about 580ft. (177m.) to 480ft. (146m.) while a depression ran across the area which although shallow at first, became more pronounced as it approached the Avon Valley. On 15th May, Capt. G.M. Paine, MVO, RN was appointed Commandant of the School and during the next five days the other members of the Staff were appointed. Of these three were officers of the Royal Navy, one was a Royal Marine and five were Army officers which, including the Commandant, made a total of ten.[26] The Commandant arrived at Upavon on 25th July to take up his duties and by the following month sufficient aircraft had arrived to enable the first course to start on 17th August.[27] Capt. Paine afterwards became Rear-Admiral Sir Godfrey Paine KCB, MVO but his final promotion followed a rather unusual course. In 1919 the Royal Air Force offered him a permanent commission with the rank of Air Vice-Marshal which he accepted but on his retirement he was placed on the Naval Retired List with the rank of Rear-Admiral.[28]

Among the earliest pupils who entered the School were several who, in later years, became some of the most eminent members of the Royal Air Force and of these two may be mentioned. Maj. H.M. Trenchard (afterwards Marshal of the Royal Air Force, Lord Trenchard, GCB, OM, GCVO, DSO) who was Assistant Commandant from 1st October 1912 until 7th August 1914 and in later years was known as 'the Father of the Royal Air Force'. The second was his contemporary Capt. J.M. Salmond (afterwards Marshal of the Royal Air Force Sir John Salmond, GCB, CMG, CVO, DSO) who was appointed an Instructor at the School soon after he had completed his course.

July and August were also periods of considerable activity at Larkhill since the War Office had decided to hold a competition there, known as the Military Aeroplane Trials, from 1st–25th August. Each machine that entered had to comply with a number of requirements in respect of such matters as take off, landing, payload and endurance.[29] Sixteen temporary hangars were erected in or near Tombs Road in order to provide additional accommodation for the twenty-four machines which had been entered, but after the competition these sheds were taken down and re-erected at Farnborough.[30] During the period of the competition, No. 3 Squadron moved to Upavon and on its return to Larkhill at the end of August, the officers were quartered in The Bustard Inn and the men at Bulford Camp.[31] They later transferred to better accommodation in the Cavalry Barracks at Netheravon where they remained until the huts and hangars at the new airfield above the village were ready[32] and to which the Squadron finally moved on 16th June 1913.

On 5th July 1912 two members of the Air Battalion, Capt. E.B. Lorraine, RE and Staff-Sergeant R.H.V. Wilson, were killed when the Nieuport monoplane in which they were flying, crashed close to the crossroads on the Stonehenge–Shrewton road (A344) a mile and a half west of Stonehenge. This was the first fatal accident on Salisbury Plain that had been caused by flying and a stone cross was erected in their memory at the crossroads which are known locally as Airman's Corner or Airman's Cross. It bears the following inscription.

TO THE MEMORY
OF
CAPTAIN LORRAINE
AND STAFF-SERGEANT WILSON
WHO WHILST FLYING ON DUTY, MET WITH
A FATAL ACCIDENT NEAR THIS SPOT
ON JULY 5TH 1912
ERECTED BY THEIR COMRADES.

Almost exactly a year later, Maj. A.W. Hewetson, RFA was killed on 17th July 1913 when he crashed in a Bristol Coanda monoplane, on the airfield at Larkhill during the test for his aviator's certificate. The exact position where his plane crashed (grid reference 139439) is marked by a stone slab surrounded by four white posts but the lettering on the stone was difficult to read in 1986. However a cross was also placed at the south-east corner of Fargo Plantation adjoining the Stonehenge–Shrewton road (A344). This is inscribed with the following words which would appear to be the same as those on the stone on the airfield site.

IN MEMORY
OF
MAJOR ALEXANDER WILLIAM HEWETSON
66TH BATTERY ROYAL FIELD ARTILLERY
WHO WAS KILLED WHILST FLYING
ON THE 17TH JULY 1913 NEAR THIS SPOT

Inevitably other fatal accidents followed, the first occurring at Upavon on 3rd October 1913 when Maj. G.C. Merrick was killed. Six months later the *Warminster Journal* when reporting on another accident at Upavon stated that 'This is the fourth aviation fatality within the last ten days' but it is not clear if all these were at the Central Flying School.[33] In August 1914 the *Journal* reported the death of two airmen at Netheravon where the Bleriot monoplane in which they were flying, crashed.[34]

As well as being of the greatest assistance for reconnaisance, aeroplanes were of the utmost value for directing artillery fire and the procedure was described in the *Salisbury Journal* when reporting on the summer exercises of 1913.

> A very interesting feature was the use of aeroplanes to direct the fire of artillery against a concealed battery ... There were three or four aeroplanes belonging to the Military Wing of the Royal Flying Corps at the exercise and though there was a fresh wind, they were almost continually at work and sent down constant and good reports of the position of the dummies. These reports were thrown down in little bags with streamers attached and orders were sent up to the flying men by cypher in the form of block letters of white canvas, pegged out or laid flat on the ground. These letters, it was found, had to be 6feet long to be read with ease. In the case of the concealed battery, an officer of the Flying Corps found the guns, reported their exact position and further reported the effect of the shrapnel. As a result there were a few hits on the target and the modest success was striking enough to make it clear that nothing should be spared to continue these experiments and to fit out each brigade of artillery with a supply of planes and fliers.[35]

In addition to establishing the Central Flying School at Upavon, it was decided to provide another aerodrome in place of Larkhill and by the autumn of 1912 the choice had fallen on Netheravon. Work began on erecting the hangars during the winter of 1912–1913 and by the following summer these and the other camp buildings were completed. As a result No. 3 Squadron was able to move from Larkhill to Netheravon on 16th June 1913 and No. 4 Squadron arrived from Farnborough two days later.[36] Before considering the development of Netheravon as a Royal Flying Corps station, it should be borne in mind that since the beginning of this century, Netheravon has been associated with the Armed Forces in three ways: through the cavalry, through small arms and through flying.

On 31st March 1898 the War Department acquired the Netheravon Estate from Sir Michael Hicks Beach, Bt. who, at that time, was Chancellor of the Exchequer. However it was not until the summer of 1904 that the Cavalry School was established at Netheravon House and an indoor riding-school and stables were built in the grounds.[37] In January 1906 the School moved temporarily to Bulford[38] but returned to Netheravon the following summer[39] where it remained until the beginning of the First World War when it 'closed on mobilization'.[40]

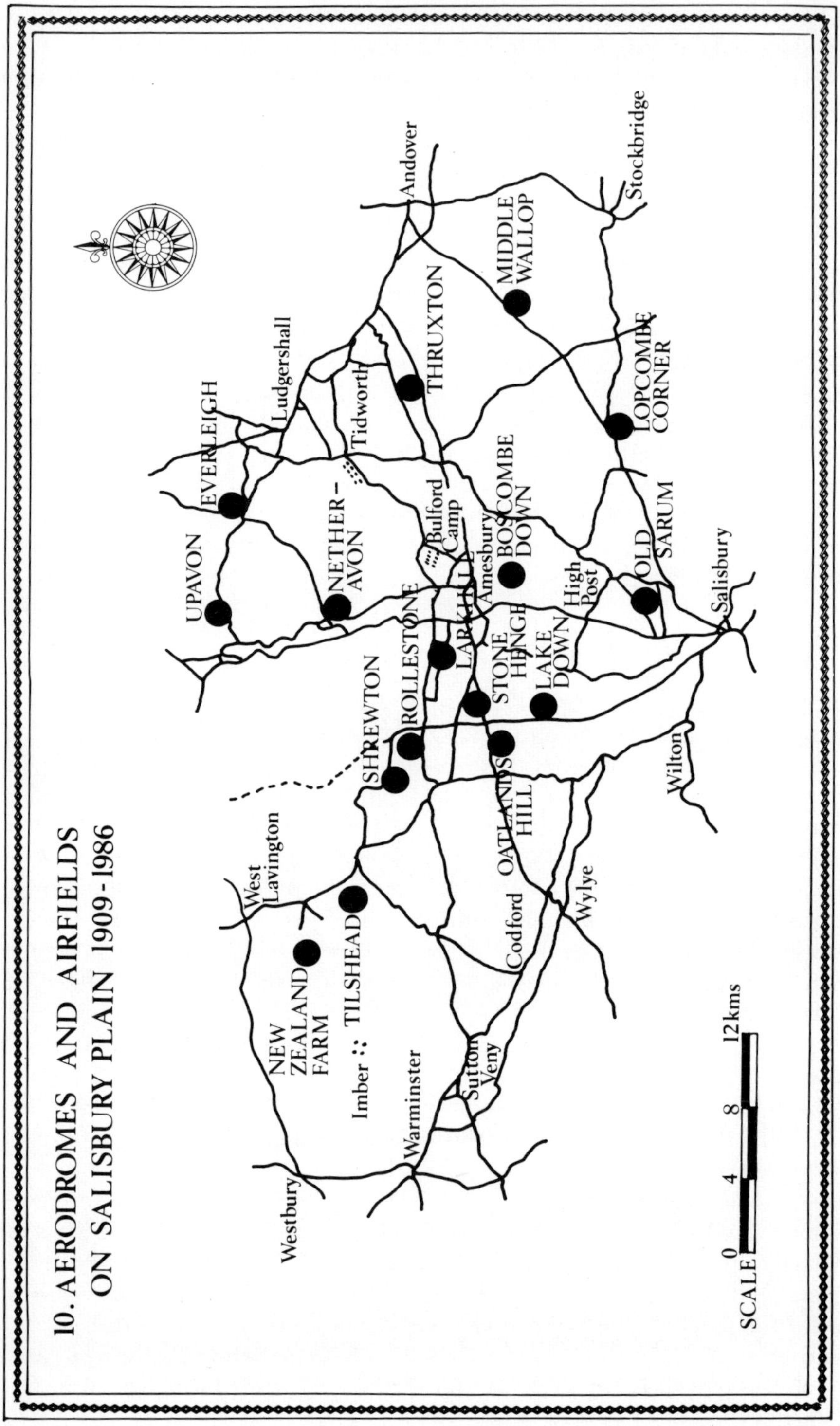

10. AERODROMES AND AIRFIELDS ON SALISBURY PLAIN 1909-1986

The School re-opened at Netheravon in October 1919[41] but in September 1922 it was stated to be 'in process of transfer to Weedon'[42] in Northamptonshire and this was completed in October when it became part of the School of Equitation.

With the departure of the Cavalry School, the accommodation was taken over as a Machine Gun School in October 1922.[43] In the Army List for August 1926, this was described as 'Small Arms School – Netheravon Wing (Vickers Machine Gun)' and it continued to be shown as such until 1939 although in later years the words 'Vickers Machine Gun' were omitted. During the Second World War the entries in successive Army Lists show it under the general heading of Small Arms Schools as 'Netheravon Wing (Netheravon)'. In July 1945 it appeared as the 'Infantry Heavy Weapons School (Netheravon)' but in 1948 it became the Support Weapons Wing of the School of Infantry.

By the early summer of 1914 there were clear signs that war could not be far away and on 2nd June the Bristol Flying School left Larkhill and moved to Brooklands.[44] Later in the month the five squadrons which were equipped with aeroplanes, that is, Nos. 2 to 6, concentrated at Netheravon where they camped under canvas. The object was to test personnel in flying, reconnaisance, tactics, aerial photography, communications and other aspects of their training[45] and the camp ended early in July. Although it had been laid down on the formation of the Royal Flying Corps that there should be a military and a naval wing, the Royal Navy had always retained a considerable degree of independence. Consequently the formation of the Royal Naval Air Service on 1st July 1914 apparently came as no great surprise and it continued its separate role until the formation of the Royal Air Force on 1st April 1918. After No. 3 Squadron had moved to Netheravon in June 1913 the airfield at Larkhill had been used mainly by the Bristol Flying School but after the School moved in June 1914 the airfield closed. When war was declared, work began on building two hutted camps on the site of the airfield and these subsequently became Larkhill Camps Nos. 2 and 3.

In December 1914 an Experimental Flight was formed at the Central Flying School of which the original staff comprised Capt. A.H.L. Soames, The King's Own Hussars and Lieut. G.M.B. Dobson, RFC who had previously been the civilian meteorologist at Upavon. Shortly afterwards they were joined by Lieut. R. Bourdillon who, before joining up, had been a don at Oxford and one of their early achievements was the production of a bomb sight which was later used in France. Sadly Soames was killed in June 1915 when a 500lb. on which he was working, blew up.[46] Among others who joined the Flight from the Universities were Lieut. H.T. Tizzard (afterwards Sir Henry Tizzard, GCB, AFC) and Lieut. F.A. Lindemann (afterwards Lord Cherwell, PC, CH) who in the Second World War, became personal assistant to Sir Winston Churchill.

During the first few months of the War, Netheravon became an extension of the Central Flying School in order to train officers who had not yet obtained their aviator's certificate but in January 1915 this responsibility was transferred to Shoreham. Netheravon then became a base for the formation of new squadrons under the command of No. 4 Wing and this arrangement continued until December 1916.[47] In May 1916, on account of Capt. Soames' death, armament testing was discontinued at Upavon and removed to Orfordness while in October, the Experimental Flight was re-named the Testing Squadron, RFC but remained at Upavon.[48] There is no doubt that flying was much enjoyed by those enthusiasts who had joined the Royal Flying Corps but for others it did not always hold the same attraction.

> The only aerodrome I remember was the Army Air Co-operation Squadron at Netheravon and we used to be given flights there, perilous indeed, in open RE8 biplanes. I remember my first flight there, sitting in an open cockpit in front of the pilot, with the propeller in the rear.[49]

In January 1917, the Testing Squadron was transferred from Upavon to Martlesham Heath some five miles to the east of Ipswich and was designated the Aeroplane Experimental Station.[50] During the course of 1917 no less than five aerodromes or airfields opened on Salisbury Plain although in some cases the buildings were not finished until 1918. These were Lake Down, Lopcombe Corner, Old Sarum, Stonehenge and Boscombe Down. Lake Down was situated between Druid's Lodge which is on the Salisbury–Devizes road (A360) and a mile and a half south of the cross roads on the A303 road, and the village of Lake near Great Durnford. This was No. 2 Training Depot Station for British and American pilots with Wing Headquarters at Druid's Lodge and until the line closed in 1920,[51] it marked the end of the Larkhill Military Railway. Lopcombe Corner was a small airfield lying between the Salisbury–Andover road (A343) and the Salisbury–Stockbridge road (A30) and operated as a Training Depot Station for Royal Flying Corps pilots.[52]

Old Sarum which was two miles north of Salisbury, occupied a site between the road from Castle Hill to Ford and the Old Roman road – the Portway – from Old Sarum to Winterbourne Gunner. It was originally used for training day bomber pilots but later became No. 11 Training Depot Station. In 1920, the School of Army Co-operation was established at Old Sarum and continued there until 1944 while in 1941 it concentrated on the training of Air Observation Post (Air OP) Squadrons. In December 1944 the School of Army Co-operation became the School of Air Support in which the Royal Navy was involved, and in May 1947 the name was changed again when it was designated the School of Land/Air Warfare. However, on 1st April 1963, it was amalgamated with the Amphibious Warfare School at Poole under the title of the Joint Warfare Establishment. On 15th

December 1971 Old Sarum was transferred from the Royal Air force to the Army and 622 Gliding School occupied the aerodrome from October 1976 until November 1978. It closed in 1979 although it has since been used occasionally as a landing ground for light aircraft.[53]

The aerodrome at Stonehenge which was used by the School of Navigation and Bomb Dropping lay astride the Wylye–Amesbury road (A303). The southern half occupied a triangular area, the sides of which were formed by the main road, Normanton Gorse and a track running northwards from Normanton Down back to the main road. The northern half which included the hangars, was on the opposite side of the main road and was served by a branch of the Larkhill railway, the eastern boundary of the site being about a quarter of a mile from the centre of Stonehenge. The landing ground lay to the north between the A303 and A344 roads and opposite the south-west corner of Fargo Plantation were other hangars. These were used by the Royal Naval Air Service and were usually referred to as the 'Handley-Page hangars' as these aircraft were normally based there. When the aerodrome closed in 1920, the buildings adjoining the A303 road were taken over by a farming enterprise and an Ordnance map of 1926 names the site as the 'Stonehenge Pedigree Stock Farm'.[54] Later in the year the site was acquired by Messrs. W.E. Chivers & Sons of Devizes and the buildings demolished in 1927 as they were considered to impair the surroundings of Stonehenge. Subsequently a bungalow and a pair of cottages which had been built between the monument and the junction of the A303 and A344 roads, were taken down for the same reason.[55]

On 1st October 1917, No. 6 Training Depot Station opened at Boscombe Down which was known as Red House Farm until 6th October when its present title was adopted.[56] The aerodrome buildings were not completed until November 1918 and in order to provide a railway connection, a branch 530yds (485m.) long was built into the site from the Newton Tony–Amesbury line. However, as an operational aerodrome, Boscombe Down was short lived since it closed for flying on 15th May 1919 and became a storage unit, finally coming to an end on 1st April 1920 when the land was returned to its owners.

For some time the future of military aviation had been receiving much thought and attention and eventually on 17th May 1916, it was announced in the House of Commons that the Air Board had been set up.[57] This was only an interim measure and on 29th November 1917 the Board was replaced by the Air Ministry.[58] It was followed on 1st April 1918, by the formation of the Royal Air Force which was brought about by amalgamation of the Royal Flying Corps and the Royal Naval Air Service. At Upavon the year 1918 was marked by an event of a very different kind – the Great Gale of 4th–5th November which destroyed seventeen aircraft and two hangars and during the course of which two Handley-Pages became airborne and landed half a mile away.

As might be expected a number of developments took place during the years which followed the First World War. In July 1919 the Netheravon Flying Training School was formed but in December its name was changed to No. 1 Flying Training School and it continued at Netheravon until 1931.[59] In 1920 the title of the Aeroplane Experimental Station at Martlesham was amended to the Aeroplane Experimental Establishment while in June 1921, Orfordness was closed and the work transferred to Martlesham. A further change in name took place on 24th March 1924 when the Aeroplane Experimental Establishment became the Aeroplane and Armament Experimental Establishment and this title was still in use in 1986. During the 1920s and 1930s, the Establishment was involved in a number of undertakings which created much public interest and these included the Schneider Trophy races, the Hendon Air Displays and several long distance flights.[60]

Although the Royal Naval Air Service had become an integral part of the Royal Air Force in April 1918, there was still a feeling in the Royal Navy that naval air operations should be entirely under its control. On the strength of the case put forward by the Navy which gained much support in Parliament and the Press, the Balfour Committee was appointed in 1923. Its principal recommendation which came into force in 1925, was to the effect that a Fleet Air Arm should be formed within the Royal Air Force which would consist of all the aircraft assigned to ships of the Royal Navy.[61] This arrangement continued until 1937 when, as a result of renewed pressure by the Admiralty, all aircraft on ships and carriers were placed under the absolute control of the Royal Navy and consequently the Fleet Air Arm became the responsibility of the Navy.[62]

By 1926 it had been decided to transfer the Central Flying School from Upavon to Wittering in Northamptonshire which is about three miles south-east of Stamford and on 7th October 1926 the School took up residence there.[63] During 1927 the site of Boscombe Down airfield was re-purchased and in the following year, work began on building a permanent aerodrome to accommodate two bomber squadrons so that the station was able to re-open in 1930.[64] After No. 1 Flying School had moved to Leuchars in 1931, 509 Flight of the Fleet Air Arm was formed at Netheravon in October 1932 and remained there until it joined *HMS Glorious* in January 1933.[65] Another link was formed between the Plain and the Fleet Air Arm in 1934 when four squadrons – Nos. 800, 801, 820 and 821 were stationed at Upavon. They left on 1st September 1935 and on the following day the Central Flying School returned to Upavon after an absence of nearly nine years.[67]

As it seemed that another war could not be long delayed, steps were taken to improve and develop air stations throughout the country and in January 1938 work began at Upavon. Further south, adjoining the main Salisbury–Andover road (A343) the construction of a new

(*Capt. D. A. Chance*)

55. 'A' Mess in July 1939 showing little change from the book-plate.

(*Defence Land Agent*)

56. Mr. A. G. Edwards of Urchfont Hill Farm herding cattle on West Down Ranges between the Wars.

(*Gunner*)

57. Visit of HRH the Prince of Wales to the Royal School of Artillery on 25th May 1979 to see the work of the Battle Group Commander's Course. From left to right: Lt.-Col. J. H. Pitman RHG/D, His Royal Highness, Col. T. D. G. Quayle and the Commandant Brig. J. D. W. Goodman.

(Imperial War Museum)

58. A 6 in. howitzer about to fire on the Larkhill ranges in November 1939. The wood immediately in front is Little Folly with part of Shrewton Folly in the right hand background.

(E. C. Booker)

59. Tarleton helmet worn by the Warminster Troop of the 'Regiment of Wiltshire Yeomanry Cavalry' in *c.* 1800. Named after Lt.-Col. Sir Banestre Tarleton, it was worn by Light Dragoons, the Royal Horse Artillery, yeomanry and some other regiments.

(*The Warminster Historical Society*)

60. Royal Engineers marching through Warminster when returning to camp. 1914.

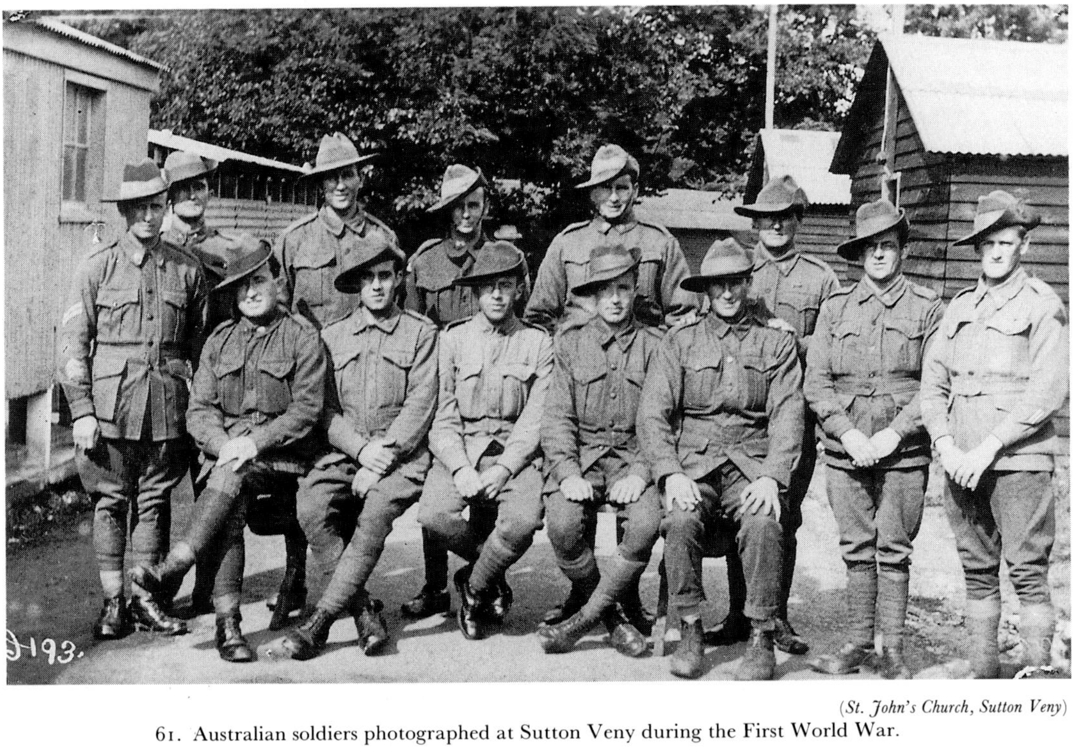

(*St. John's Church, Sutton Veny*)

61. Australian soldiers photographed at Sutton Veny during the First World War.

(*P. Cole*)

62. The Australian badge on the hillside above Foxhole Bottom, Codford St. Mary north of the Warminster–Salisbury road.

(*Michael Stratton*)

63. Cartoon of Codford published in 1916 by W.H. Malcolm, Tobacconist, Trowbridge and Codford.

(*The Tank Museum*)

64. A Royal Tank Corps camp at Warminster probably in 1932, showing the effect of tanks on a grass camp site in wet weather.

(*Imperial War Museum*)

65. The Royal Tank Corps Band at Warminster probably for the opening of the new barracks on 23rd September 1938. On the left is Bandmaster D.J. Plater and in the centre, Trumpet Major Low.

(*Imperial War Museum*)
66. A tank hangar at Warminster in 1938. On the left are Vickers Medium Tanks and on the right Light Tanks, MK. VIB.

(*Imperial War Museum*)
67. Gen. Eisenhower and Air Chief Marshal Sir Arthur Tedder watching members of the US 33rd Armored Division firing on a miniature range operated from a Sherman tank. Warminster 26th February 1944.

(T. L. Fuller & Sons)

68. The hangars of the British and Colonial Aeroplane Co. situated in what is now Wood Road Larkhill.

(T. L. Fuller & Sons)

69. The funeral procession of Capt. E. B. Lorraine who was killed in a flying accident on 5th July 1912, approaching Bulford Station.

aerodrome started a mile to the east of the village of Middle Wallop. Later that year HM King George VI honoured the Central Flying School with a visit on 5th May. In 1939 a naval advanced Flying Training School for Fleet Air Arm pilots was formed at Netheravon which continued to operate there until April 1942[68] and after war was declared on 3rd September, the Aeroplane and Armament Experimental Establishment moved from Martleshal Heath to Boscombe Down.[69]

Before considering some of the events and changes which were brought about by the Second World War, the closing stages in the employment of balloons should be recorded. Although the balloon had been replaced by the aeroplane it was still used for artillery observation and in July 1916, the Royal Flying Corps were allocated 50 acres (20 hectares) opposite Rollestone Bake Farm, near Larkhill, on which to establish a balloon school. This was concerned with the training of personnel in the use of kite balloons and was designated No. 1 Balloon School. Two large hangars were built on the site which, according to Ashworth, were erected in 1932[70] and these were still a prominent feature of the surroundings in 1986. Adjoining these hangars is a small wood known as Rollestone Clump and part of the eastern perimeter was planted in the form of the letter 'E'. It is probable that this was done in order that balloons could be moored in the spaces between the rows of trees which formed the horizontal bars of the 'letter' and iron rings set in concrete, to which the mooring ropes were attached, were still on the site in 1986.

In 1916 an 'outstation' of the School was set up at Tilshead and the report of the Lands and Buildings Reconstruction Committee of 1918 records, under the heading of 'Schools' a 'Kite Balloon School' at both Rollestone and Tilshead.[71] The report also refers to an aerodrome at Rollestone which was presumably part of the School and was the area from which the balloons were flown. Some of the activities of the Tilshead Balloon School have been recalled by Mr. G.H. Griffith who spent much of his early life in the village.

> While the balloons were flown from many sites in the area, the equipment was located in a field on the right hand side of the road leading from Tilshead village to West Down Camp, immediately above the gardens of the last house on this road (possibly grid reference 038479). The balloons were sometimes navigated through the village to the flying sites by a number of men hanging on to the balloon guy ropes for dear life. A mobile winch was not apparently used for this purpose although the balloons appeared to be fully inflated. German prisoners of war were employed in the maintenance of these balloons ... and the gas was held in containers similar to the present day oxygen cylinders. There were normally two balloons at Tilshead, each complete with baskets and each basket having an encased parachute hanging from it. These parachutes were in containers ... which remained attached

> to the basket, the parachute being pulled out in an emergency descent by the weight of the falling observer.[73]

Mr. K.E. Chant whose family also lived in Tilshead, can recollect as a small boy, another occasion when a balloon was being manoeuvred through the village.

> I remember watching a balloon being deflated just outside our shop as the men tried to get it under some telegraph wires and some local people dowsed their fires lest there should be an explosion – all great fun to us.[74]

During the First War, 105 Squadron RFC operated from a temporary airfield which was situated some 800yds (730m.) east of the Tilshead–West Lavington road in the area of grid reference 025490.[75] From 1925 until 1939 Army Co-operation squadrons used a landing ground about a mile to the west of Tilshead village (reference 021478) during the annual manoeuvres.[76]

By 1920, the School at Rollestone had been re-named the Royal Air Force School of Balloon Training and in 1925 it was largely concerned with the observation of artillery fire which was practiced in conjunction with the School of Artillery at Larkhill. Aircraft were used as well as balloons and the procedure is described in the following account by Mr. H. Powpey who was serving in 1st Heavy Brigade RA which carried out part of its annual training at Larkhill.

> Two batteries trained with the 9.2in. and two with the 8in. howitzers and we usually camped under canvas due south of Knighton Flagstaff. From 1927 to 1934 I was GPOA (Gun Position Officer's Assistant) at many firings with the 9.2s. The observation was either from balloons which were provided from the School at Bustard Camp or from aircraft. In the case of the first named, the orders were by telephone to the ground and were controlled by Battery officers but when aircraft were used RAF personnel received the orders by radio. The aircraft used were Bristol Bulldogs and Wallaces which flew on figure of eight courses passing over the target area and the battery position, at the top and bottom of the loops. The order to fire was given as the plane was approaching the intersection of the figure, the flying speed being about 90mph. The line of fire ran across Slay Down and most shoots were at a range of about 12,000–13,500yds (10,970–12,350m.) the 8in. hows covering a much wider zone than the bedded down 9.2s.[77]

It was not long before further changes were made in the name of the School: in 1929 it became the Larkhill Kite Balloon Station; in 1932, the RAF Balloon Centre and in 1936, No. 2 Balloon Training Unit. Up to 1935 observation from balloons was included in courses for Battery Commanders and Young Officers and also on Gunnery Staff Courses and the following describes an ascent in 1933.

> We had our lecture on the technique of airshoots, where an RAF plane observed our fire, and the following day I was No. 4 on a 6in. how. on the

ranges. I was ordered to double to the rear to the balloon winch vehicle – a Peerless lorry, I believe. As I arrived there, the balloon was just coming down to the ground. A fellow YO, looking very green in the face tottered out of the basket and I was ordered to enter and ascend.

My knowledge of airshoot procedure was almost nil but my companion in the basket, a young, wingless and charming RAF officer, seemed to have quite a good grasp. Using the telephone line which ran down our tether cable, from some 700feet up, together we ranged and went to fire for effect on some white screens which were out of sight from the ground and IG level. Naturally, we reported that fire for effect had been bang on target.

Up in that balloon, on a bright sunny day, with the larks singing and the voices clearly coming up from the gun position was one of my most pleasant memories of Larkhill and the YO's Course.[78]

At the same time, there were others who would have much preferred to have stayed on the ground.

On some exercise in 1938 I was told to go up in a balloon and 'keep observation'. I was terrified as I ascended in a laundry basket with an RAF chap. When we got up, I peered over the edge of the basket and found the country was oscillating through about 60°. One look and I was sick and the RAF chap waved a red flag and we were pulled down. I can still remember the sight of my breakfast slowly floating away down wind. The BC who was crazy about the air went up and would not come down. The CO wanted to talk to him and I said he was not available because he was in a balloon. I think he thought that I was taking the mickey out of him.[79]

In 1939 the RAF Anti-Gas School took over Rollestone Camp and the Balloon Training Unit was disbanded although in May 1940, the German Army used balloons for artillery observation in France and Belgium with considerable success.[80]

Although the buildings were not completed until the end of 1940, Middle Wallop became operational on 15th April of that year[81] in spite of the fact that some personnel were under canvas, cooking was done on field kitchens and the officers were quartered at RAF Andover.[82] The new aerodrome which was smaller than it is today, was first occupied by No.15 Service Flying Training School but this moved at the beginning of June when Fighter Command took over the station. Between June 1940 and July 1941, fourteen enemy bombing attacks were made on the aerodrome, the most serious being on 14th August when the casualties amounted to six killed and fifteen injured and a hanger was demolished. In an attempt to reduce these attacks, a 'decoy' station was built near Houghton some four miles to the south.[83]

As had been the case in 1917, a number of small military airfields were established in 1940 which comprised High Post, Everleigh, New Zealand Farm and Shrewton. Of these, High Post was unique in that it was a private airfield which had been opened by the Wiltshire Light Aeroplane and Country Club in 1931. It closed in September 1939 but on 6th June 1940 it was taken over by 112 Squadron of the Royal

Canadian Air Force. After its main factory in Southampton which produced Spitfires had been bombed in September 1940, the airfield was taken over by Vickers Supermarine in 1941 for the erection and flight testing of aircraft. After the War, the Wiltshire Flying Club returned to High Post in April 1946 but the airfield closed when the Club moved to Thruxton in June 1947.[84]

The airfield at Everleigh which lay to the north of the village and about three miles east of RAF Upavon, opened in November 1941 but was closed at the end of 1945 during which time it was used by the Central Flying School and also as a Maintenance Unit dispersal area.[85] A little over two miles due east of St. Joan à Gores Cross (on the Tilshead–West Lavington road) is New Zealand Farm which was established in October 1940 when it was used by RAF Upavon for training pilots in night flying. It was closed and transferred to the War Department on 17th December 1943 which was the date, referred to in Chapter 11, when the village of Imber was evacuated on account of the expansion of the Warminster ranges. After the end of the War, New Zealand Farm became an outstation for Boscombe Down.[86]

Shrewton lay within an area bounded by Greenlands Farm camp, The Bustard, Shrewton Farm and Elston Hill. It came into operation in July 1940 as a relief landing ground for Netheravon and in 1941 was used for night flying and later by Army Co-operation units. With the arrival of gliders at Netheravon in 1942, Shrewton became a landing practice ground but it was little used after the end of 1944.[87] Although hardly qualifying for the designation of 'airfield', an area (grid reference 088439) lying between the A360 and B3086 roads was requisitioned in September 1939. This was used by the RAF Anti-Gas School at Rollestone in order to carry out simulated gas attacks from the air, on personnel attending courses. When the School moved from Rollestone in October 1945, the landing ground was taken over by aircraft of 657 Squadron (Air OP) but it closed in July 1946 and the land was transferred from the Royal Air Force to the Army.[88]

At Tilshead, the landing ground which lay a mile to the west of the village was brought into use again in June 1940 and early in July, 225 Squadron which was equipped with Lysanders, moved into a tented camp adjoining the airfield. This was known as Tilshead Lodge Camp and when the weather deteriorated, Tilshead Lodge was taken over but by the end of 1941 the Squadron had moved to Thruxton.[89] During this period the village was surrounded by camps with Lodge Camp to the west, the Airborne Camp (grid reference 036487) half a mile to the north, West Down artillery camp to the east and 'C' Camp (reference 039467) a mile to the south. The use of airborne troops by the German Army no doubt hastened the formation of the Army Air Corps which comprised the Glider Pilot Regiment and the Parachute Regiment, in January 1942.[90] The first of these two regiments moved to Airborne Camp in January and towards the end of the month it was joined by

C Company of the 2nd Battalion Parachute Regiment which had been selected to carry out the raid on Bruneval but after this operation, it returned to its base in Derbyshire. In June 1942, 21st Independent Parachute Company arrived at Airborne Camp but was transferred to Newcome Lines, Larkhill, in the following October while the Glider Pilot Regiment moved to Fargo Camp during September, 1943, their place being taken by an armoured regiment. When the United States Forces arrived on the Plain in the autumn of 1943 they were allocated Lodge Camp and 'C' Camp.[91]

Another small airfield was opened in June 1941 at Oatlands Hill between Hill Farm, Winterbourne Stoke, and Long Barrow Cross Roads and this was bounded on the west, north and east sides by the road leading to Hill Farm, the A303 road and the A360 road respectively. It was concerned with training pilots for the Army Co-operation Command but was little used after August 1944 and finally closed on 13th May 1946.[92] A considerably larger base was established at Thruxton which began to operate in August 1941 and it was from here that the Bruneval raid was carried out. It was later occupied by the Glider Pilots Exercise Unit and in 1944 by the United States Army Air Force. After the War it was leased to the Wiltshire School of Flying which had vacated High Post and it is now also used as a motor racing circuit.[95]

A number of events marked the year 1941: on 8th May HM King George VI visited Middle Wallop; a High Altitude Flight was formed at Boscombe Down[94] and in November, Netheravon was chosen as the base of the new Glider Exercise Squadron.

This was allotted the number 296 and in December the first of the new airborne forces arrived including 297 Parachute Exercise Squadron.[95] Consequently in January 1942, Netheravon was transferred from flying training to the Army Co-operation Command while on 19th January, No.38 Wing was formed there in order to control 296 (Glider) Squadron and 297 (Parachute) Squadron and to provide the necessary air requirements for the Airborne Division.[96] Headquarters for the 6th Airborne Division were established at Syrencote House, a mile and three-quarters to the south of Netheravon and a plaque commemorating this event, was later placed besides the front door.

In April 1942 the naval advanced Flying Training School which had been at Netheravon since the beginning of the War was closed, as arrangements had been made to train some of the Fleet Air Arm pilots in the United States.[97] Later in the month, the Prime Minister, Mr. Winston Churchill, attended a demonstration of parachuting and the use of gliders and on 21st May Their Majesties The King and Queen visited Netheravon where they inspected detachments from 296 and 297 Squadrons. On 27th July the Station was paid a visit of a very different kind when the first and only enemy attack occurred. The officers' quarters and those occupied by the WRAF were machine gunned but there were no casualties and only slight damage. At

Upavon, fundamental changes were taking place and when, on 12th March 1942, the Empire Central Flying School was established at Hullavington, Upavon was re-named No.7 Flying Instructors School.[98] In the field of research, an Intensive Flying Development Flight was formed at Boscombe Down in order to assess the long term reliability of aircraft. This was followed in May 1943 by the establishment of a Test Pilots' Training School.[99] Netheravon was honoured by another visit of HM The King on 1st April 1943 while in October 38 Wing became 38 Group. Finally the year was brought to a dramatic close when on 31st December, four American Flying Fortress bombers made an emergency landing on the airfield two of which had been hit.

On 10th January 1944, 67 Tactical Reconnaissance Group of the United States Army Air Force took over Middle Wallop but on 1st August the aerodrome was returned to the RAF.[100] Less than three weeks before the invasion of Europe, Their Majesties The King and Queen accompanied by HRH The Princess Elizabeth visited Netheravon on 19th May. In attendance were Air Marshal Sir Trafford Leigh-Mallory, KCB, DSO Air Officer Commanding-in-Chief of the Allied Expeditionary Air Force; Air Vice-Marshal L.N. Hollinghurst, CB, OBE, DFC Air Officer Commanding 38 Group; Maj.-Gen. F.A.M. Browning (afterwards Lt.-Gen. Sir Frederick Browning, GCVO, KBE, CB, DSO) commanding the British Airborne Corps and Maj.-Gen. R.N. Gale (afterwards Gen. Sir Richard Gale, GCB, KBE, DSO, MC) commanding 6th British Airborne Division.

On 15th February 1945 Middle Wallop was taken over by the Royal Navy and commissioned as *HMS Flycatcher*, its task being to organize and establish Mobile Naval Air Operation Bases (MONABs) but on 10th April 1946, Middle Wallop became a RAF station once again.[101] During 1945 the Empire Test Pilots' School moved from Boscombe Down to Cranfield and Farnborough but returned in 1968[102] while Netheravon became an important air transit centre. With the end of the War the inevitable changes began to take place and these included the re-forming of the Central Flying School on 7th May 1946, not at Upavon but at Little Rissington between Burford and Stow-on-the-Wold.[103] Upavon, on the other hand, saw the arrival of Headquarters 38 Group and later the Headquarter Transport Command, but this arrangement did not last for long and 38 Group was disbanded on 1st February 1951[104] although Transport Command continued at Upavon. As an aerodrome, Netheravon had closed by the end of 1950 and on 1st December it was taken over as an RAF Police Training Centre which included a Police Dog School and a Police Driving School. However, during June and July 1951, flights of 662, 663 and 664 Auxiliary Air OP Squadrons visited Netheravon.

In January 1948, 657 Air OP Squadron RAF, followed by 227 (Air OP) Operational Conversion Unit arrived at Middle Wallop and this was an indication of the role that the station was to play in the future.

During 1950 the Conversion Unit became the Air OP School which was responsible for training Army pilots in elementary flying and was later known as the Light Aircraft School.[105] In spite of the fact that the Army Air Corps which was formed in February 1942, had been disbanded in 1950, by 1957 there was an increasing need for light aircraft and helicopters in the Army. Consequently in July of that year the present Army Air Corps came into being under Army Order 82,[106] and on 1st September 1957 the Light Aircraft School became the Army Air Corps Centre. At the same time it was appreciated that the assistance of the Royal Air Force would be needed after the transfer in September 1957, so it was not until 1st October 1958 that Middle Wallop became an entirely Army station and the RAF Headquarters were disbanded.[107] When, in 1954, the Weighbridge Hangar was erected at Boscombe Down, it made a considerable impact on the local skyline. Its vast size allows work to be carried out on the largest aircraft, under cover and at the same time provides extremely accurate weighing facilities.[108]

Although 38 Group had been disbanded in February 1951, it was re-formed again at Upavon in 1960 but it was soon realized that the Station was not large enough to accommodate this Group, as well as Transport Command. The difficulty was solved by transferring 38 Group to Odiham in May 1960 and erecting a new building in 1962 for Headquarters Transport Command which was re-named Air Support Command on 1st August 1967.[109] The year 1962 marked the fiftieth anniversary of the establishment of the Central Flying School on 19th June 1912 and a special display of flying and static exhibits was held at Upavon on 16th June 1962, to celebrate the event. The static display was visited by HRH The Duke of Edinburgh and at the final flypast by No. 1 Squadron RAF, the salute was taken by Marshal of the Royal Air Force Sir John Salmond who had been on the first course to be held at the School in 1912.[110]

In 1962 the Police Training College moved from Netheravon and the Lower Airfield Camp became an Army transit area but the Upper Camp fell into a state of disrepair which was made worse by the activities of vandals. However, during 1963 the Army Free Fall Parachute Association took over part of the Upper Airfield Camp and this subsequently became the Joint Services Parachute Centre.[111] The year 1964 saw a considerable development in Army Aviation and in August the headquarters of 651 Squadron moved from Middle Wallop to Netheravon where it was divided in order to form two Army Aviation Headquarters.[112] This was followed by a period of considerable activity in the reorganization and movement of units and at Middle Wallop the School of Army Aviation was set up in August 1965 and the title of the Army Air Corps Centre changed to the Army Aviation Centre.

By 1971 work at Boscombe Down had increased to such an extent that it was necessary to organize the Establishment into eight Divisions, namely: Armament, Engineering, Flying, Navigation and Radio, Performance, Photographic, Technical Services and Trials Management.[113] At Middle Wallop the name of the Army Aviation Centre reverted to that of the Army Air Corps Centre in January 1973 while the designation of the School of Army Aviation was discontinued in March of that year.[114] On 1st September 1972, Air Support Command was combined with Strike Command and on the same day 46 Group was re-formed and established its headquarters at Upavon. A further change occurred on 10th November 1975 when 38 and 46 Groups were combined under the title of 38 Group which returned to Upavon to form the largest Group in the Royal Air Force.[115] Finally, on 18th November 1983, 38 Group was amalgamated with 1 Group which took over at Upavon and was still there in 1986.

When the Central Flying School opened in 1912 only a few buildings had been completed and work continued over the next three years. In 1986 there were still twenty of the original buildings in existence although the use to which some of them had been put, had changed. The earliest are five single officers quarters built in 1913 on the east side of York Road, four of which are located to the north of the Officers' Mess and one to the south. However, during 1914 considerable progress was made and fourteen buildings were completed. Of these, eight comprised the Sergeants' Mess, the Regimental Institute (now the Avon Club), five small barrack blocks and a building for NCO's quarters situated between the Sergeants' Mess and the Avon Club. The remaining six buildings consisted of three houses for Warrant Officers in what is now Devon Road, one of which still bears the date '1914', a squash court near the Officers' Mess, an observatory which has been converted into a post office and bank and the Regimental Office.[116] This last building is now The Trenchard Building Museum which was opened by Viscount Trenchard, MC on 9th September 1983. It includes the Commandant's Office which was used by Capt. G.M. Paine and this contains a number of personal items such as his binoculars. Displays in other rooms cover the early days of the Central Flying School, some aspects of the work of Marshal of the Royal Air Force Lord Trenchard and the development of Upavon from the 1920s to the 1980s.

By far the largest of the original buildings is the Officers' Mess in York Road and while the date over the entrance is 1915, this records the year in which it was completed although it was in use by 1914. It is the oldest officers' mess in the Royal Air Force which is still in service but at one time this claim was contested by Netheravon. However, it was finally established that the Upavon mess was completed a few weeks before its opposite number at Netheravon and it was consequently able to uphold its claim to seniority.[117] A plan of the School area dated 30th December 1917 which was issued by Salisbury Plain District show two lines of 'aeroplane sheds'. One of these lines comprising eleven sheds, ran parallel to and south of, the Upavon–Ludgershall road and occupied the site of the present day hangars. The other, numbering seven sheds which were used by the Experimental Flight, were located along the eastern boundary of the Camp at right angles to the Ludgershall road.[118]

Although an Ordnance map published in 1925 marks a Roman Catholic Church opposite and on the north side of the present NAAFI, it does not show one for the Church of England but only a 'Church Hall' opposite the Sergeants' Mess.[119] However the Ordnance map of 1974 names this building as 'St. Peters Church (C of E)' and also marks a Catholic Church of St. Thomas More between York Road and the most westerly of the three hangars.[120] It would appear that St. Peter's did not become a church, as opposed to a building in which services were held, until 1956 since a plaque in the Church bears the following inscription.

This Church was Dedicated to the Glory of God
and named St. Peter's at a service on
28 October 1956.

The bell was originally a ship's bell in *HMS Centaur*, an aircraft carrier of 22,000 tons which was completed in 1953, placed in reserve in 1966 and scrapped in 1971. Since 1985 the Church has been used by all denominations.

The first married quarters for airmen at Upavon were built in Beverley Crescent in 1932 with additional quarters being provided in Pembroke and Oxford Roads after 1937. The officers' married quarters were built during two main periods, those in Comet Avenue and Hastings Avenue between 1930 and 1950 and those in Britannia Avenue and Anson Avenue during the 1950s. Of the remaining station buildings the following may be mentioned: the two westerly hangars were erected in 1926 and that on the east sometime later, while Station Headquarters and the Headquarters of 1 Group were built in 1962 an extension being added in 1968.[121]

When Netheravon aerodrome was laid out, it was divided into two sections which are a little less than half a mile apart, the larger one being known as Lower Airfield Camp and the smaller as Upper Airfield Camp. The road which joins these two sites runs through the middle of the Lower Camp and most of the buildings immediately to the south of this road are part of the original development. Their external design is most unusual for military buildings and consists of white panels bordered by vertical and horizontal timber bonding strips painted black which puts one in mind of the old black and white half-timbered buildings of Warwickshire. These buildings form two adjacent hollow squares, the one on the west which has the Officers' Mess in the centre, is flanked by single officers' quarters. In the square on the east side, the Warrant Officers' and Sergeants' Mess which was previously the Regimental Institute, forms the centre piece with a line of accommodation huts on the remaining two sides. Further to the east is a hangar which dates from about 1925. The Officers' Mess is of particular interest not only in view of its age and construction but also on account of the original carved wings of the Royal Flying Corps which have been placed over the main entrance.

On the north side of the road and close to the Camp entrance is the Quartermaster's Office and stores which was built in 1919 and has since been enlarged. Adjoining this is the old Church of England church which, like the Roman Catholic church has now been closed while beyond it is the old camp hospital, at present occupied by the Headquarters of the Army Air Corps in the United Kingdom. Further to the east are the Soldiers' Dining Hall and the NAAFI both of which are of modern construction while behind these are a number of huts for the accommodation of single soldiers. The original married soldiers quarters were built in Choulston Close but in 1986 these were empty and due for demolition, new quarters having been built behind them in 1952. The officers' married quarters which have been built at various times since 1925, are located in Kerby Avenue. An Ordnance map of 1925 shows a line of eight hangars at the Upper Airfield Camp but in 1986 only one of these remained and this was occupied by the Joint Services Parachute Centre.[122] In the meantime two new hangars had been built to replace those that had been demolished while another which was marked on the 1925 map, was in use by the Army Fire Services in 1986. This map also shows a number of small ancillary buildings behind the line of hangars but in 1986, only six of these were still standing.

Due to the fact that Middle Wallop was built between 1938 and 1940, many of the original buildings are still in use although a number were added during the period from 1951 to 1967. Amongst those which comprised the original air station the principal ones are the Officers' Mess, the Warrant Officers' and Sergeants' Mess, the present Junior Ranks Club, the barrack blocks and the five hangars. Steps were taken to camouflage the buildings in May 1940 and it was still possible to see traces of this in 1986, more particularly on parts of the Officers' Mess and barrack blocks. During 1940, seventeen quarters were built in Falaise Road for married officers and sixty-seven in Willow Way and Sycamore Crescent for married airmen. From 1951 to 1955 six more married officers' quarters were built at the north-west end of Falaise Road and these were followed by another three in Ipoh Close between 1955 and 1957.[123]

A major rebuilding programme began in 1966 when the opportunity was taken to demolish a number of old buildings, blast walls and airfield defences. Of the new buildings, the largest is the Stockwell Hall which was named after Gen. Sir Hugh Stockwell, KCB, KBE, DSO who was the first Colonel Commandant of the Army Air Corps from 1957 to 1963.[124] This provides accommodation for the Ground Instructional

Wing, the Aircraft Engineering Training Wing and the Aircraft Servicing Branch, REME. Another forty-nine officers' married quarters were built in and around Taiping Road and some hundred and twenty quarters for married soldiers on the north side of the Camp. In the early 1950s the Royal Air Force operations room was converted into the Church of St. Michael (Church of England) and although it was subsequently altered to some extent, it is still possible to recognize its original role. The Church is now used by all denominations except Roman Catholics, who have their own Church in a building which was dedicated to St. Assisi in 1980.[125]

When the Museum of Army Flying was opened by HRH Prince Michael of Kent on 7th July 1984, the work which had begun nearly thirty-five years earlier, was brought to a successful conclusion. In 1948 a collection of historical items which had been formed at RAF Andover, was moved to Middle Wallop but as the size of the collection increased so did the difficulties in finding a suitable building in which to display it. The problem was finally overcome when the present Museum which adjoins the Andover–Salisbury road (A343) some 500yds (460m.) south of the hangars, was built.[126]

CHAPTER THIRTEEN

Old Roads and Railways of The Plain

In the early days tracks led the way from one village or town to another and at first these were used by travellers who were mounted or on foot, as well as by pack animals. As the use of wheeled carts and wagons increased, it was found these tracks which were now being referred to as roads, soon became impassable during spells of wet weather. If a heavily laden wagon became stuck, it was not unusual to leave it where it was while the ground dried out which might not take place until the following summer. The movement of wagons was sometimes so slow and unreliable that pack animals were often preferred for carrying merchandise that could be divided up into small light loads.

During 1663 the first turnpike roads were established which were so called on account of the barriers or 'turnpikes', consisting of spikes fixed to a rotating bar, that were placed across the road. Their purpose was to prevent travellers from passing until they had paid a toll which was used to meet the cost of repairing the road. The Ordnance Survey map of 1817 shows two turnpikes at Amesbury, one at Countess and the other near the cemetery on the old road to the west; one at Stapleford, another at Knook and four at Warminster, three of which were on the present A36, A350 and A362 roads respectively.[1] However it was only after new methods of road construction had been introduced by such men as Metcalfe, Telford and McAdam that conditions began to improve. The course of the old roads over the Plain were in some cases, indicated by milestones, signposts or boundary stones and the position of some of these are shown on both the old and the modern maps. Andrews and Dury's Map of Wiltshire published in 1773, marks a number of 'direction posts' including one at Long Barrow Cross Roads between Stonehenge and Winterbourne Stoke and another to the south-east of Knighton Long Barrow near the south-west corner of the present RA Mess at Larkhill. This must have been an important point as the post is also depicted by Cary on his Map of Wiltshire of 1801.

One of the oldest roads on the Plain is the section of the Ridge Way which is shown on the Ordnance Survey as running from St. Joan à Gores Cross, over Urchfont Hill and along the escarpment to the top of Redhorn Hill.[2] From there it continued eastwards over Wilsford Hill and then turned north-eastwards to the village of Charlton. It is believed to have formed part of a system of ancient roadways which crossed England from East Anglia to the south-west. Another road of the same period was the Harrow Way which linked Steeple Langford with Berwick St. James and after crossing the Salisbury road near Druid's Lodge, continued over Normanton Down, almost to Stonehenge. It then turned east passing a little to the north of Amesbury and followed a course which was very similar to the present A303 road, as far as the turning to Cholderton.

Four roads ran northwards from Salisbury across the Plain and probably the most important of these was the Devizes road which followed what is now the A360 as far as Airman's Cross. From there it continued past Rollestone Camp to The Bustard and after a further two and a half miles, reached the junction with the old road to Market Lavington which was still marked by a signpost in 1986. This road which branched off to the north-west, crossed Gore Down and Church Hill before reaching Lavington. By tradition it was here that in 1785, Jarvis Matchan was confronted by a spectre which unnerved him to such an extent that he confessed to the murder of Drummer Benjamin Jones. Matchan had joined the Royal Navy in 1778, had deserted twice and after the second occasion, entered the Army from which he also deserted. Eventually, being destitute, he enlisted in the 49th Regiment of Foot and on 19th August 1780 was ordered with Drummer Jones, to draw subsistence money from the recruiting officer, some six miles from Huntingdon where they were stationed. During their return journey, Matchan became drunk, murdered and robbed the drummer and absconded to the Navy. Five years later he was finally discharged at Plymouth and with a shipmate – John Shepherd – set out to walk to London and during the course of their journey made use of the road to Devizes.

In the middle of the afternoon, as they were approaching the turning to Market Lavington, Shepherd pointed to something about a hundred yards to their right. When Matchan went to investigate, he saw that it was the upper half of a woman's body, much deformed, that appeared to shake its head at him. After returning to where Shepherd was waiting, they both went to see this apparition which continued to behave in the same way but soon disappeared leaving only a large stone in its place. As they continued their journey some of the stones in the road seemed to turn towards Matchan and on reaching an inn, he thought that he saw a vision of Our Lord with the murdered drummer. As a result, Matchan confessed to the murder, was convicted at his trial at Huntingdon and was executed on 2nd August 1786. A pamphlet

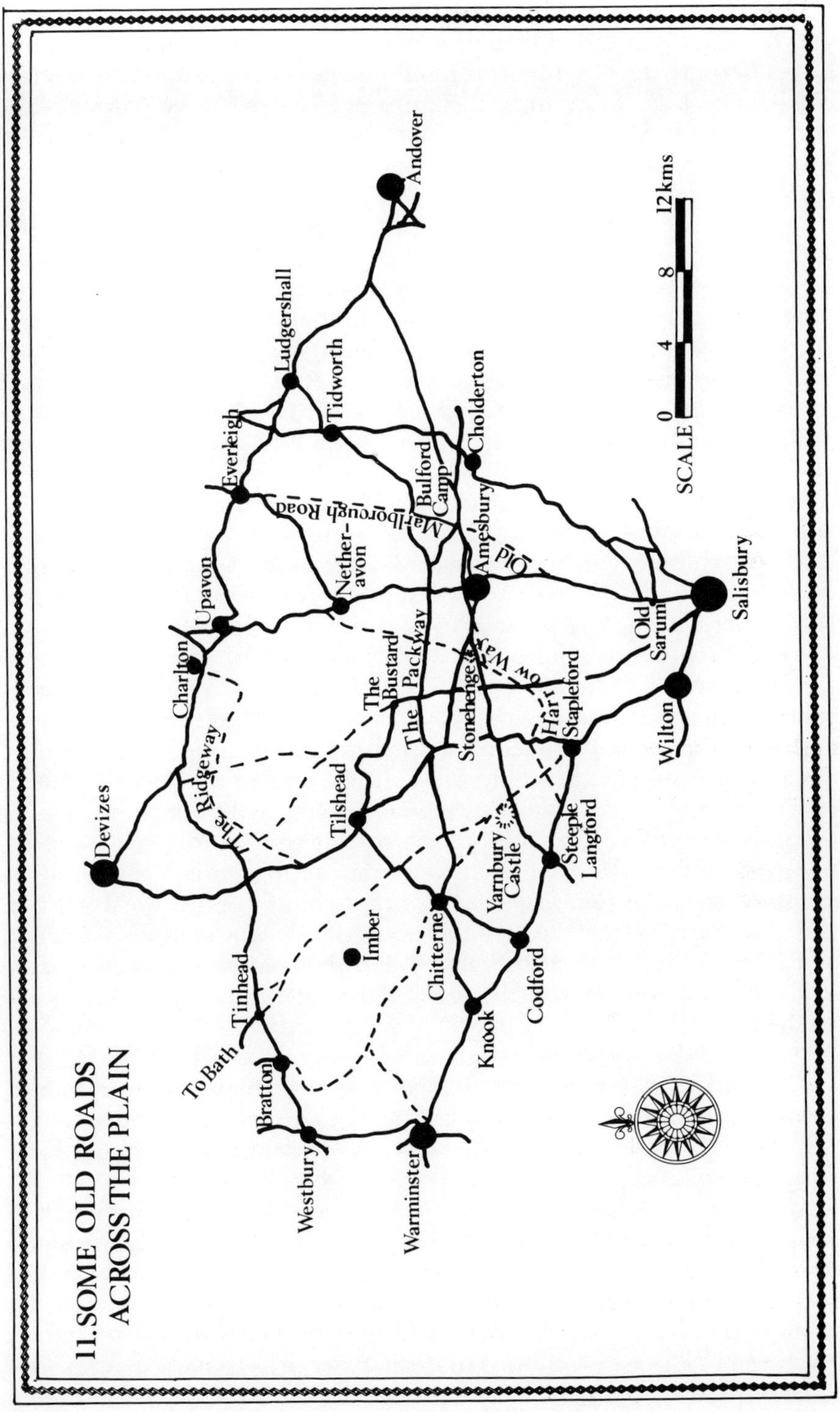

11. SOME OLD ROADS ACROSS THE PLAIN

describing these events was published shortly afterwards[3] and a poem entitled 'The Dead Drummer' was included in *The Ingoldsby Legends* which were written by the Rev. R.H. Barham under the assumed name of Thomas Ingoldsby. This contains the following lines which will be appreciated by those who have found themselves in a similar situation.

It is a very sad thing to be caught in the rain
When night's coming on upon Salisbury Plain.[4]

The Devizes road proceeded northwards passing some 300yds (275m.) to the west of Ell Barrow and after crossing the old road from Netheravon to Easterton, reached the Ridge Way at the top of Redhorn Hill and continued to Devizes. The Ordnance map of 1958[5] marks four milestones between Rollestone Camp and Redhorn which recorded the distance to Salisbury and Devizes. In 1911 the ranges at Larkhill were extended towards Tilshead, and the adjoining area to the west of the Devizes road became known as the West Down ranges.[6] As a result the use of the road by the general public was considered to be too dangerous and it was closed during the day. At the same time the road from Shrewton and Tilshead to West Lavington was virtually re-built and in 1912 the use of the old road by the public was further restricted and is now closed when firing is taking place on the ranges.[7]

To the east, the road from Salisbury to Marlborough provided another link across the Plain from south to north. On leaving Salisbury it passed to the east of Old Sarum and followed the course of the present road to Amesbury (A345) until it reached a point a mile to the north of High Post. From here it continued in almost a straight line to the north-east through Boscombe Down crossing the main road (A303) at grid reference 178422. It then followed the present road to Bulford Camp passing through the centre by way of Marlborough Road and Sling Road to Sheepbridge Road. After crossing this road, it headed northwards for about 900yds (825m.) to a ford over Nine Mile River and then passed through Milston Firs and along the western edge of Brigmerston Plantation. The old road then crossed Haxton Down and Weather Hill before reaching Everleigh and so to Marlborough. On the current Ordnance maps its course is still inscribed as 'The Old Marlborough Road'.

Two roads from Salisbury crossed the western portion of the Plain which lies between the Shrewton-West Lavington road (A360) and the Wylye Valley road (A36). These were the Salisbury–Bath road and the Salisbury–Warminster and Bratton road and in the early stages they both followed the same route along the A36 road to Serrington, a quarter of a mile to the west of Stapleford. Here they turned to the north-west across Steeple Langford Cow Down, past a milestone which recorded 'Bath 28, Sarum 8' and over the A303 road to Yarnbury Castle. Some 500yds (460m.) beyond the northern perimeter of the

Castle, the two roads separated, the Bath road bearing to the north and the Warminster road continuing in a north-westerly direction.

The road to Bath crossed the Shrewton–Chitterne road at grid reference 021434 and after passing Chitterne Barn, joined the road which runs from Tilshead to Chitterne for about 180yds (165m.) before continuing northwards over Chitterne and Chapperton Downs. After following a line over the road from Imber to St. Joan à Gore Cross, it passed to the east of New Zealand Farm and over Stoke Hill before entering the village of Edington. The Ordnance map published in 1958 marks eight milestones on the road from Yarnbury Castle to Edington which record the varying distances to 'Bath' and 'Sarum'.[8]

To the north of Yarnbury Castle, the Warminster road crossed part of Deptford Down and joined the Shrewton–Chitterne road (A344) at a bend some 300yds (275m.) west of New House Farm. It then followed the course of this road through the southern part of the village which is named on the Ordnance map of 1817 as Chiltern St. Mary but which later became Chitterne St. Mary. At the western end of the village where the present road turns sharply to the south, the old road branched away to the north-west and after passing Bowl's Barrow, reached a cross roads at grid reference 315479 which is sometimes referred to as Imber Clump. At this point the road to Warminster turned to the south-west down Sack Hill and into the town while the principal road continued northwards to another road junction at reference 900501. Here the Westbury road branched away to the north-west and the Bratton road carried on northways to that village. Between Chitterne and Warminster, three stones are marked on the Ordnance map of 1958 which record the mileages to 'Warminster' and 'Sarum'.[9]

Apart from the road from Amesbury to Wylye (A303) which Cary names as the 'Wincaunton and Taunton Rd', the other principal link across the Plain from east to west, was the road from Amesbury to Rollestone, Shrewton, Chitterne and Heytesbury. At its junction with the Wylye valley road (A36), Cary inscribes it as leading to the 'Bruton or Frome and Wells Rd.'[10]

Although probably of less importance than the roads which have already been described, two others may be mentioned. The first ran from approximately grid reference 118418 on the A303 road, past Stonehenge and across the Devizes road to The Cursus. From there it continued in almost a straight line through what is now Larkhill Garrison to reference 133470. It then proceeded in a north-easterly direction to join the Amesbury–Upavon road (A345) a little to the north of Netheravon at reference 143494. The second road which is still in use, is The Packway that runs from Bulford village to the Stonehenge Inn and through Larkhill to Rollestone cross roads and Shrewton.

Today it is difficult to imagine what the Plain was like before motor vehicles and the Army made their mark upon it – and its roads. At the end of the last century, Dr. W.M. Childs who later became the first Vice-Chancellor of Reading University, visited the Plain and his son, Brig. C. Childs CBE, has recalled the following incident which provides a glimpse of the past.

> I remember my father telling me that in 1896 or 1897, he crossed the Plain on a cycle. It was a lonely and very beautiful place with hardly any traffic on the tracks that served for roads. Somewhere around Shrewton, a shepherd came running across to him to pass the time of day, saying that my father was the first living person he had seen for the past two days.
>
> I must add that from my memory, most of the roads were not much better than tracks when I first arrived on the Plain in 1919.[11]

The first military railway was built during the Crimean War (1854–1856) and ran from the town and harbour of Balaclava to a supply point behind the British lines. With a route length of five and three-quarter miles and two branches each of about half a mile, it was begun in January 1855 and completed in May of that year.[12] Although Russell referred to it rather facetiously in his despatches as 'the Grand Crimean Railway', it proved to be of the utmost value in moving ammunition and stores.[13]

As the nineteenth century evolved, this new form of transport became increasingly important and by 1882, Salisbury Plain was encircled by railways. These are referred to below under the name of the parent company with the year in which they came into operation.

Great Western Railway
- 1851 Westbury – Warminster.
- 1856 Warminster – Salisbury.
- 1862 Savernake – Pewsey – Devizes.

London & South Western Railway
- 1857 Salisbury – Andover.

Midland & South Western Junction Railway
- 1882 Andover – Ludgershall – Savernake.

In 1900 the Great Western built a line from Patney and Chirton some four miles east of Devizes, to Westbury in order to provide a more direct route to the West of England.

However there were no direct railway connections with any of the localities which were rapidly developing into important military centres. Trains carrying troops bound for Bulford or Larkhill, stopped at Porton station and after detraining, units were faced with a march of six miles over country roads to Bulford or eight to Larkhill.[14] The nearest station to Tidworth was Ludgershall and although it was only two miles away, all stores, rations and ammunition had to be conveyed by GS wagons from the station to the barracks. The road out of Tidworth climbed 145ft. (44m.) in the course of a mile before it reached the summit opposite Pickpit Hill, and this added to the difficulties for horse drawn wagons especially in snow or frosty weather. Consequently one of the first matters to receive attention after the War Office had begun to acquire land on the Plain, was the provision of railway facilities.

The Great Western Railway lost no time in putting forward a scheme and at the first meeting of the newly appointed War Office Salisbury Plain Committee, held on 11th May 1897, it was agreed to examine it. The Company proposed the construction of a standard gauge railway which was referred to as the Pewsey and Salisbury Light Railway. It was intended that it should leave the Great Western main line a little to the west of Pewsey Station and after following the valley of the River Avon, would join the Warminster–Salisbury branch of the Great Western at Bemerton, now a part of Salisbury. Nine stations were planned and of these, five were to be built at Upavon, Enford, Netheravon, Durrington and 'Stonehenge and Amesbury' respectively. The last named was to be sited close to the road from Amesbury to Stonehenge at grid reference 140418 or thereabouts.[16]

At the next meeting of the Committee, a fortnight later, it was agreed that not only would the proposed line be of no value to the War Department, but it would in fact interfere with its interests and accordingly should be opposed 'tooth and nail'. It was also noted that

> The South Western were prepared to make a line from Grately, outside W.D. property, to Lavington, which would be much more suitable from our point of view.[17]

As already noted in Chapter 3, the reason why members of the Committee opposed the building of this railway was that, in their view, it would seriously interfere with the use of the Avon for practising river crossings. However this did not deter the Great Western Railway from pursuing its objective and an inquiry was held at Salisbury on 25th October 1897. The Committee felt so strongly on this matter that it was decided to refer it to Parliament for a final decision.[18] Fortunately no such action proved necessary as the Committee was informed on 20th April 1898 that the Board of Trade had refused to grant any powers to construct the railway on land belonging to the War Department and that was the end of the matter.[19]

On 24th May 1897 the Committee had been informed that the London & South Western were prepared to build a line to Lavington and six months later a draft order of the proposed 'Amesbury and Military Camp Light Railway' was laid before it. This was unanimously approved and the view was expressed that there was little doubt that the Company 'would be ready to extend the line from Shrewton to Tilshead'.[20] At the same time, the question of acquiring more land to the south of Larkhill for training purposes was under consideration and the Committee felt that the final decision on this would have a bearing on the extension of the railway from Amesbury to Shrewton. It was also thought that it would be 'an act of vandalism to run the railway near to Stonehenge' and it was agreed to ask the London & South Western if the line could be diverted to the south, that is further away from Stonehenge.[21] In reply the Company stated that the course of the line could not be altered except at 'greatly increased expense'. At the same time, favourable consideration would be given to 'a suggestion from the War Department to abandon that portion of the approved line which lies to the west of the proposed Amesbury station' subject to the payment of such expenses as had been incurred.[22] To this the Committee agreed and as it was informed on 24th January 1901 that 'there was no immediate prospect of funds becoming available for any large additional purchases',[23] no further action was taken as regards the acquisition of the land or the extension of the railway beyond Amesbury.

During these negotiations, the Midland & South Western Junction Railway had been busy preparing alternative proposals, no doubt in anticipation – and hope – that the Great Western's plan for the Pewsey and Salisbury Light Railway would be unacceptable. The first of these schemes was for a line from Ludgershall to Salisbury but at a meeting on 23rd September 1897, the Committee decided that

> the proposed line would not suit the purposes of the War Department in any way, and that the other line proposed by the South Western Railway was far preferable.[24]

Not to be discouraged, the Company submitted a new plan for a railway from Ludgershall to Amesbury to be known as the 'Ludgershall and Military Camps Light Railway' but was promptly told that the War Department disapproved of the scheme and would oppose it.[25] Although it was stated that the line would run 'through the eastern side of the War Department's property' it would appear, in the light of a later minute that this was not so. The Committee's rather brusque decision was later modified at a meeting held on 11th March 1898 when it was decided that there would be no objection 'as the proposed line ran entirely outside the boundaries of the War Department's property'. At the same time, the Committee pointed out that this railway

> must run in with the proposed line to be constructed by the London & South Western Railway from Grately to Shrewton.[26]

In the meantime the London & South Western and some of the landowners through whose property the line would have passed, decided to oppose its construction and nothing more was heard of it.[27]

The first railway to be built on the Plain with the primary object of serving a military centre, was the line from Ludgershall to Tidworth which was constructed for the War Office by Henry Lovatt Ltd., of

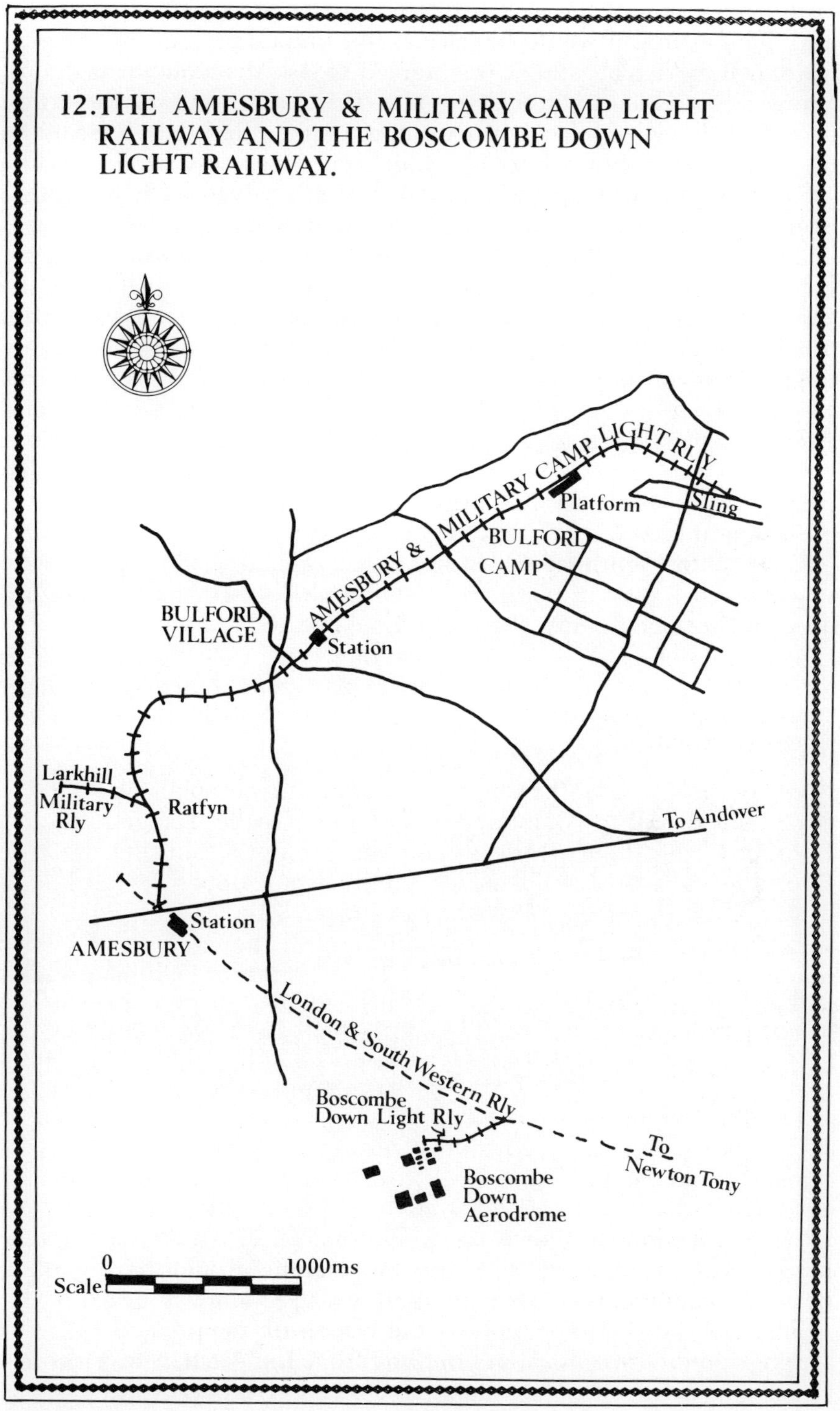
12.THE AMESBURY & MILITARY CAMP LIGHT RAILWAY AND THE BOSCOMBE DOWN LIGHT RAILWAY.
AMESBURY & MILITARY CAMP LIGHT RLY
Platform
Sling
BULFORD CAMP
BULFORD VILLAGE
Station
Larkhill Military Rly
Ratfyn
To Andover
Station
AMESBURY
London & South Western Rly
Boscombe Down Light Rly
To Newton Tony
Boscombe Down Aerodrome
0
1000ms
Scale

Wolverhampton who also built the barracks. The minutes of the Salisbury Plain Committee do not refer to this branch presumably because it was laid by the War Department on land that it already owned. The line was completed by 8th July 1901 when it was used in connection with the annual manoeuvres and for transporting the men and materials that were required in building the barracks. These constructional workers were accommodated in a camp of corrugated iron huts at Brimstone Bottom between Tidworth and Ludgershall which was known locally as 'Tin Town'. In April 1903 the contractors were granted permission to lay a siding at the camp in order to facilitate the movement of the work force to and from the barracks but on completion of the contract it was removed.[28] At first the line was worked by the Midland & South Western Junction Railway under a temporary agreement with the War Department but this was followed in February 1903 by an arrangment for the next fourteen years which was later extended.[29]

On leaving Ludgershall station, the Tidworth line turned to the south-west and at first ran parallel to the Tidworth road (A3026). After passing to the south of Tidworth House Farm and Windmill Hill Plantation it curved in a broad sweep to the south, ran through a deep cutting, crossed the Tidworth–Ludgershall road by a bridge and entered the station. The total length of the branch was 2 miles 33 chains or 2.41 miles and it opened for War Department traffic on 21st May 1902, for public goods traffic on 1st July and for public passenger traffic on 1st October of that year. At Tidworth there were two platforms: that on the west side on which the station buildings were situated was about 390ft. (119m.) long and was used by the general public. On the east was the platform which was provided for troop movements and was designed so as to afford ample space being about 828ft. (250m.) in length and 40ft. (12m.) in width.

Traffic on the branch proved to be very heavy which was not surprising in view of the demands made upon it. In addition to the movement of ordnance stores it was necessary to provide a regular supply of rations and these included some items in bulk. Once a fortnight a special meat train arrived at Tidworth while three wagon loads of flour were despatched every week, from Avonmouth.[30] Related to this was the need for coal on which cooking and heating depended and large quantities were delivered at regular intervals, each of the eight barracks having its own coal yard. The transport of horses by rail also gave rise to a considerable amount of traffic on the branch. While complete units travelled by train to and from Tidworth, a large number of service personnel travelled on the line as individuals including those going on leave, courses or when posted to other units. As a result, the annual receipts at Tidworth were more than the combined total from all the other stations on the Midland & South Western and it was accordingly recognized as the 'senior' station on the railway.[31] As

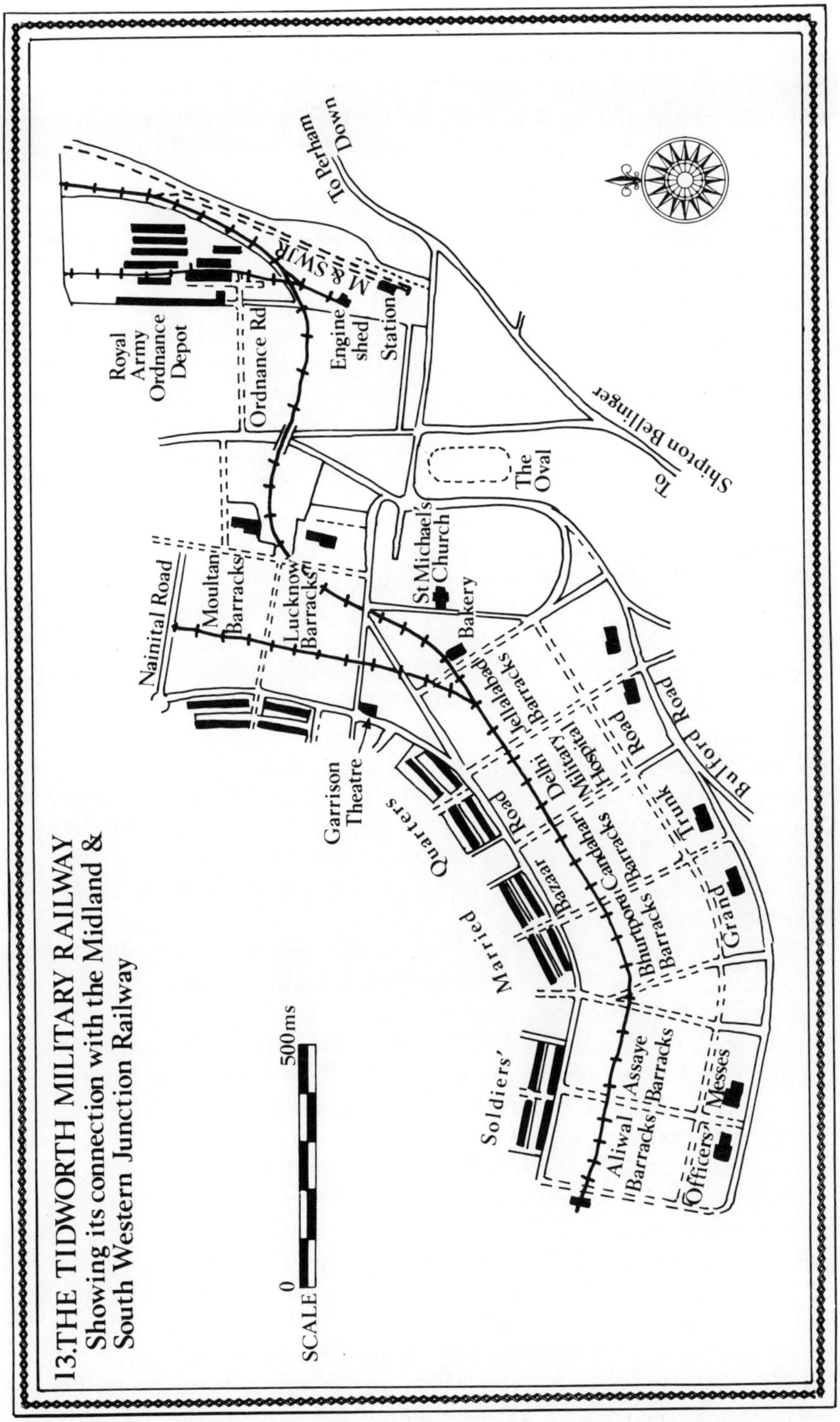
13. THE TIDWORTH MILITARY RAILWAY
Showing its connection with the Midland & South Western Junction Railway
SCALE
0
500ms
To Perham Down
M & SWJR
Royal Army Ordnance Depot
Ordnance Rd
Engine shed
Station
To Shipton Bellinger
The Oval
St Michael's Church
Bakery
Moultan Barracks
Lucknow Barracks
Nainital Road
Jellalabad Barracks
Delhi
Military Hospital
Road
Bulford Road
Garrison Theatre
Quarters
Married
Road
Bazaar
Candahar Barracks
Trunk
Bhurtpore Barracks
Grand
Soldiers'
Assaye Barracks
Aliwal Barracks
Messes
Officers'

peacetime conditions returned after the First World War, traffic gradually declined but it increased again very considerably during the Second War, especially prior to the invasion of Europe in 1944.

Under the Railways Act 1921 which came into effect on 1st January 1923, the Midland & South Western was amalgamated with the Great Western Railway and when nationalization of the railways took place on 1st January 1948, the Great Western became the Western Region of British Railways. However, it was not long before there were signs that it was unlikely that the Tidworth branch would survive for any length of time and on 19th September 1955, public passenger services were withdrawn. On 25th November public goods traffic came to an end and three days later the operation of the line was taken over by the Royal Engineers. This arrangement continued for almost eight years and although the last troop train used the station on 25th May 1962, the line remained open until 5th November 1963 when the final goods train was run.[32] The track was lifted during the winter of 1963–1964 and by 1968 the whole of the station area had been built over.

However the line from Ludgershall to Red Post Junction (about a mile and a half west of Andover) which had been singled in 1960,[33] was retained by the Ministry of Defence together with half a mile of the Tidworth branch. From this, a line was laid in 1943 into what is now the Armoured Vehicle Sub-Depot at Ludgershall, details of which will be found in Chapter 7. The goods yard was also in use in 1986 although the track layout had been modified to a large extent. The distance from Red Post Junction to the bridge which carries the Tidworth road (A3026) over the line at Ludgershall is about 5.87 miles and from this bridge to the buffer stops in the Armoured Vehicle Sub-Depot is a further 1.27 miles giving a total route length of 7.14 miles.

Although it is not clear when the Tidworth Military Railway was built, there can be little doubt that the track was laid in conjunction with the Tidworth–Ludgershall branch and was completed by the same date, namely 8th July 1901. The fact that the branch line and the Military Railway were both built by the same contractors – Henry Lovatt Ltd. – would support this. In effect, the line began at the buffer stops (grid reference 241493) which stood abreast of the original northern boundary of what was the Royal Army Ordnance Depot. From there it followed the east and south-east boundaries of the Depot and after bearing due west, crossed Pennings Road (A338) at its junction with Meerut Road, by means of a bridge. The line then turned in a wide sweep to the south, over the adjoining sports grounds and, passing between Lucknow and Mooltan Officers' Messes, crossed St. George's Road and on through Lucknow Barracks. After passing over Lowa Road, between St. Patrick's and St. Michael's Avenues, it entered a cutting and on emerging, continued between the Church of England Institute and the Garrison Bakery. About 120yds (109m.) beyond Kirkee Road, a branch from Lucknow and Mooltan Barracks joined the line. From Kirkee Road the track ran in a straight line through Jellalabad, Delhi and Candahar Barracks and having passed Cabul Road, turned west through Bhurtpore, Assaye and Aliwal terminating 35yds beyond the boundary road on the west side of Aliwal Barracks.

The branch to Lucknow and Mooltan which turned northwards in Jellalabad Barracks, at grid reference 230483, crossed Kirkee Road and proceeded between the old Presbyterian Church and the Church of England Institute. It then ran in a straight line over St. Patrick's Avenue, passing immediately to the north of the present Garrison Headquarters in Lowa Road, and on through Lucknow Barracks. Beyond Margha Road it continued by way of Mooltan Barracks to a point about 30yds to the north of Nainital Road and 210yds (192m.) west of St. George's Road, where it ended. The line was single track from Tidworth Station until it reached Kirkee Road but from there to Aliwal it was doubled with crossovers provided in Bhurtpore and Delhi Barracks. The Mooltan branch became double track immediately after it had crossed St. Patrick's Avenue and continued as such until it was a short distance from Nainital Road, a crossover being provided in Mooltan Barracks.

In the case of Aliwal and Assaye, Bhurtpore and Candahar and Lucknow and Mooltan, the coal yards were built in pairs to serve the adjoining barracks, with a loop line running through each pair. As regards Delhi and Jellalabad, the yards were separated and a short siding was provided in both cases. The electricity generating station which was built in Candahar Barracks, in 1913, was also served by a siding while the double track which extended as far as the bakery, enabled wagons loaded with flour to be discharged with the minimum obstruction to other traffic. Although part of the Royal Army Ordnance Depot which adjoined Tidworth station had been built during the First World War, it was not initially provided with a railway connection. An Ordnance map to the scale of 1:20,000 which had been issued in about 1918–1919 and republished to a scale of 1:25,000 by the War Office in 1931[34] does not mark any sidings in the Depot but a map revised in 1923[35] shows that they had been laid by that date.

The route length of the line from the dead end at Tidworth to Aliwal Barracks was approximatley 3,390yds (3,100m.) or 1.90 miles while the branch from Jellalabad to Mooltan Barracks was 793yds (725m.) or .45 miles, giving a total of 2.35 miles. The doubled section of the line, together with the sidings and loops in the barracks amounted to another 2,724 yds (2,491m.) or 1.54 miles of track while the sidings in the Ordnance Depot including the access spur added a further 1,600yds (1,462m.) or .90 miles. Thus the total length of actual track was 8,507yds (7,778m.) or 4.79 miles. The line was operated by members of the civilian staff of the War Department although other arrangements may have been made during the First World War. Since the early 1930's two locomotives had been in use, one being a 0–6–0 saddle

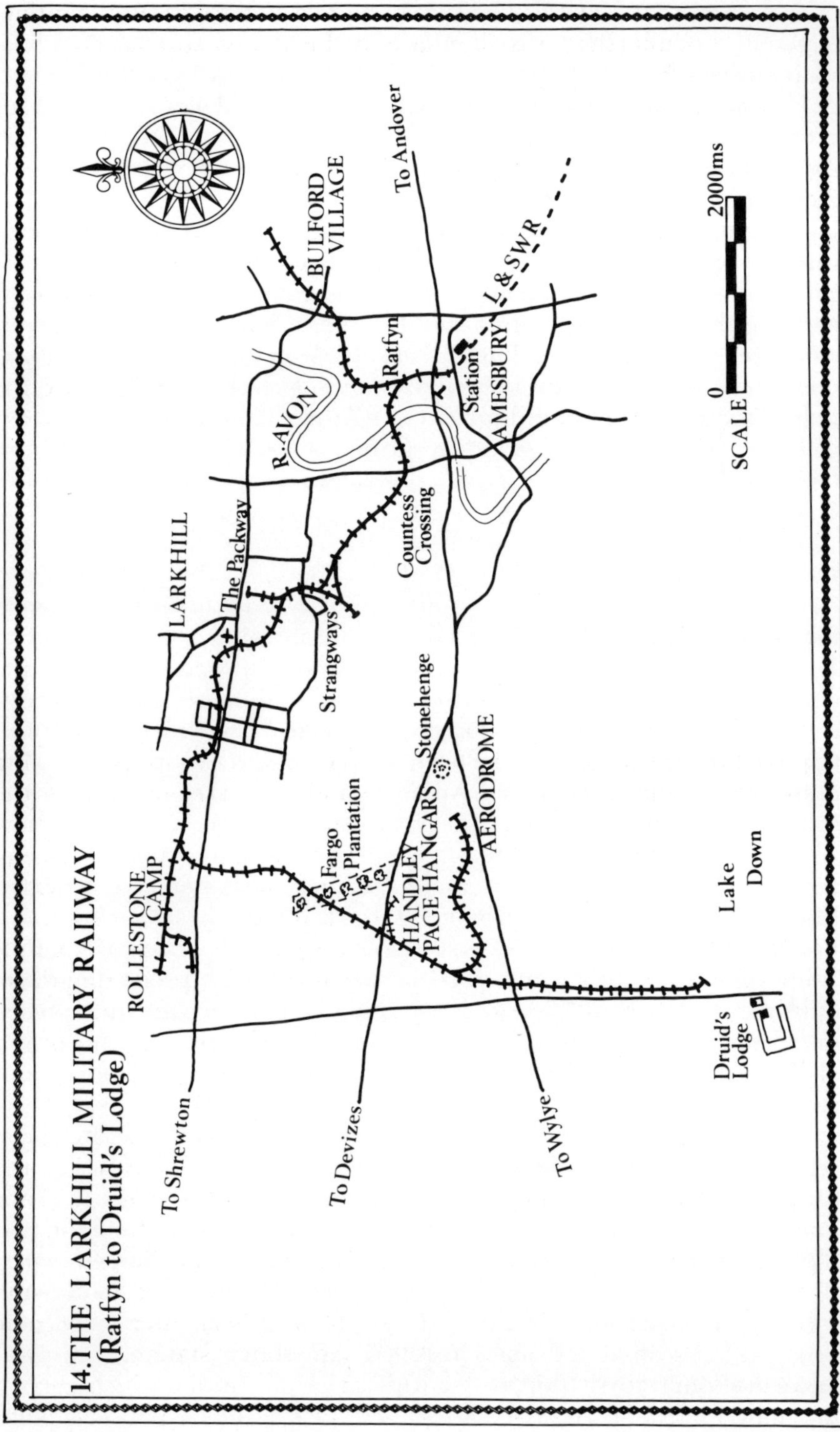

14. THE LARKHILL MILITARY RAILWAY
(Ratfyn to Druid's Lodge)

tank named 'Betty' while the other, named 'Molly', was a Sentinel locomotive of unorthodox design built by the manufacturers of Sentinel steam lorries.[36]

It is not clear when the Tidworth Military Railway closed but an Ordnance map published in 1961 which had been revised in 1957,[37] shows that the whole of the line that lay to the west of the Ordnance Depot had been lifted by that year. The track bed between Kirkee Road and Aliwal and from Jellalabad to Mooltan has been converted into two tarred roads which can be identified, since unlike other roads in the Garrison, they have not been named. A footpath now runs through the cutting between Lowa Road and Kirkee Road which can still be easily recognized. There is also some evidence of the embankment, between Calne House and the pumping station, which raised the line to the height of the bridge over Pennings Road. The only section of track which could be found in 1985 was a length which had been fixed in the concrete floor of the old electricity generating station in Candahar Barracks.

The Amesbury and Military Camp Light Railway which ran from the main Salisbury–Andover line of the London & South Western to Arnesbury, was built by the firm of J. T. Firbank of London and opened for public traffic on 2nd June 1902, with an intermediate station at Newton Tony. The initial arrangements which were made for the actual connection of the branch with the main line, did not prove satisfactory and in 1904 major improvements were carried out. On approaching the main line, the branch divided into two spurs one being eastwards in the direction of Andover and the other curving to the south-west towards Salisbury. In the case of the first, a separate track was laid besides the main line as far as Grateley station – a distance of two and a half miles – where a connection was made. This provided for trains travelling to and from Amesbury and Andover but for trains bound for Salisbury the junction consisted of two components. A simple connection with the up line met the needs of trains proceeding from Salisbury to Amesbury, but for those going to Salisbury the track passed under the main line which it joined a short distance beyond. The object of this was to enable Amesbury trains to cross the main line without causing delays.

The point at which the branch joined the Salisbury line, has been referred to as Newton Tony Junction and Allington Junction but the Ordnance map of *c.*1918 gives two names.[38] The point where the spur line diverged towards Andover is shown as Newton Tony Junction and the site where the Salisbury spur joined the main line is named Amesbury Junction. From the main line the branch proceeded to Newton Tony station and after passing the site of what was later to become Boscombe Down aerodrome, reached Amesbury station. This was located about 170yds (155m.) south-east of the bridge which carried the main road from Andover to the west before the by-pass was built,

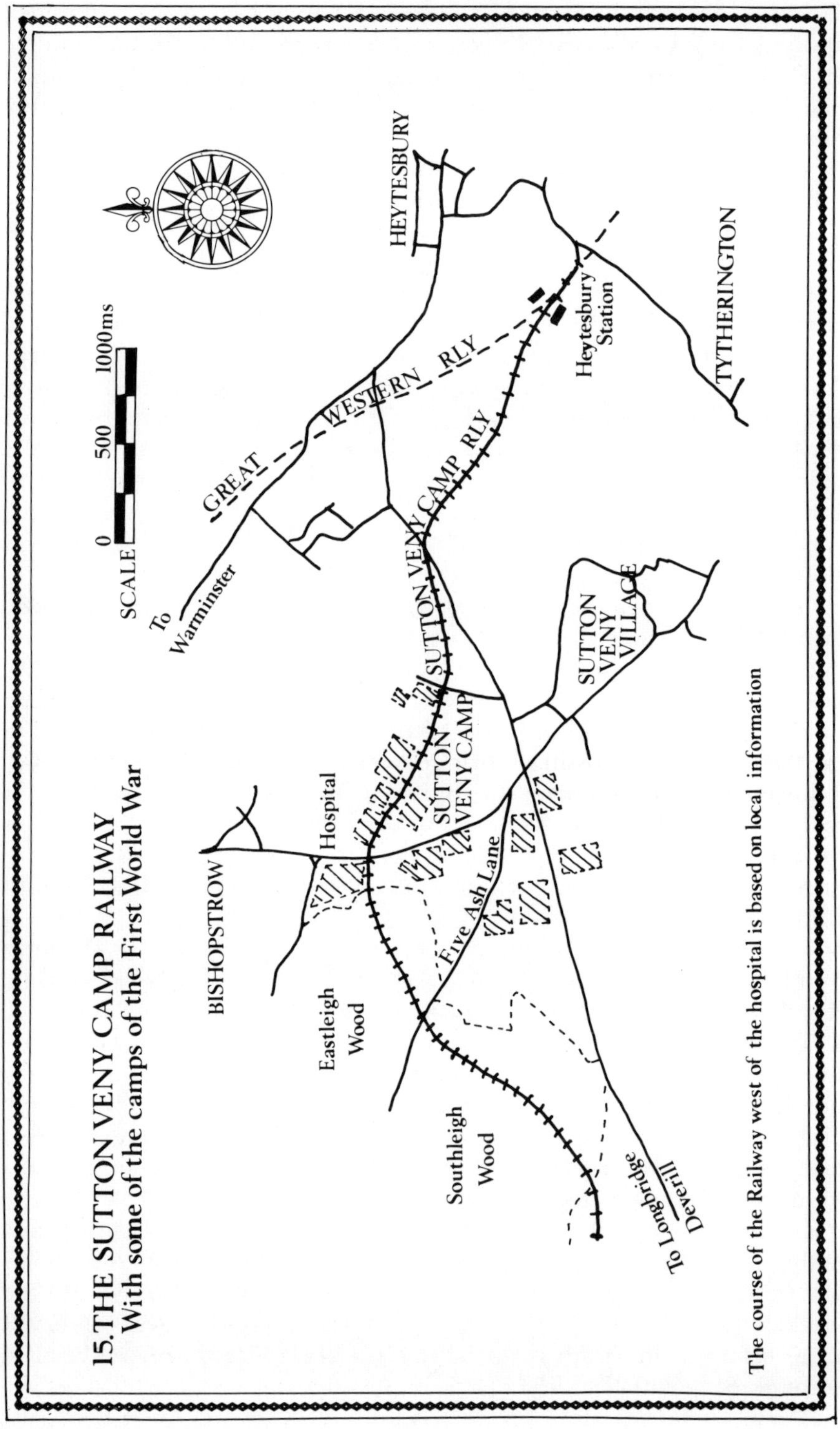

15. THE SUTTON VENY CAMP RAILWAY
With some of the camps of the First World War

The course of the Railway west of the hospital is based on local information the grid reference of the station buildings being 164419. As mentioned earlier in this chapter, it was originally intended to continue the line to Shrewton and possibly to Tilshead and Lavington. By the time it had been decided to terminate the line at Amesbury,[39] work on the Shrewton extension had already begun and a cutting about 350 yds (320m.) in length had been excavated beyond the station and this was subsequently used for accommodation sidings.

The length of the branch from the main line to the road bridge at the northern end of the station yard was approximately 4.71 miles but there were also quite extensive sidings adjoining the station in addition to those that had been laid on the site of the discarded extension to Shrewton. It has been said that before and during the First World War Amesbury was, for its size, one of the busiest stations in the country, but when the line was extended to Bulford Camp traffic probably declined although the branch as a whole no doubt continued to operate to its full capacity. After 1946 passenger traffic began to fall off and on 25th June 1952 public services were withdrawn although goods traffic continued until 4th March 1963. Shortly afterwards the branch was closed and the track was lifted in July 1965.

On 1st June 1906, almost exactly four years after the Amesbury branch had opened, the extension to Bulford Camp came into operation. This single line left Amesbury station at a point immediately beyond the bridge carrying the old Amesbury–Andover road and then turned north past Ratfyn. Within half a mile or so, the track curved to the east and after crossing a bridge over the Bulford–Amesbury road, followed by another over the Bulford–Durrington road (A3028), it reached Bulford station. Double track was installed at the station so as to allow trains to pass and a goods shed, cattle pen and sidings were provided. The line continued in a north-easterly direction to Bulford Camp where a long open platform had been built to which access was obtained from the western end of Horne Road. A siding that ran parallel with the branch extended from a coal yard (between Haig Road and Wing Road), to the eastern end of the platform – a distance of about 645yds (590m.). The track then turned eastwards in a wide sweep and after crossing Marlborough Road, terminated on the north side of Sling Plantation approximately at grid reference 192441. A short distance beyond the road a platform was built with a loop on the north side to enable a locomotive to run round a train and a siding on the south side, adjoining the platform.[40]

The route length from Amesbury to Bulford station was approximately 1.35 miles and from there to Sling a further 1.51 miles or a total of 2.86 miles excluding loops and sidings. The line was closed and the track lifted at the same time as the Amesbury line. There were still some traces of the railway to be seen in 1986, mostly in the form of embankments although the platform at Bulford Camp and parts of the platform at Sling could also be recognized. After the line closed, the

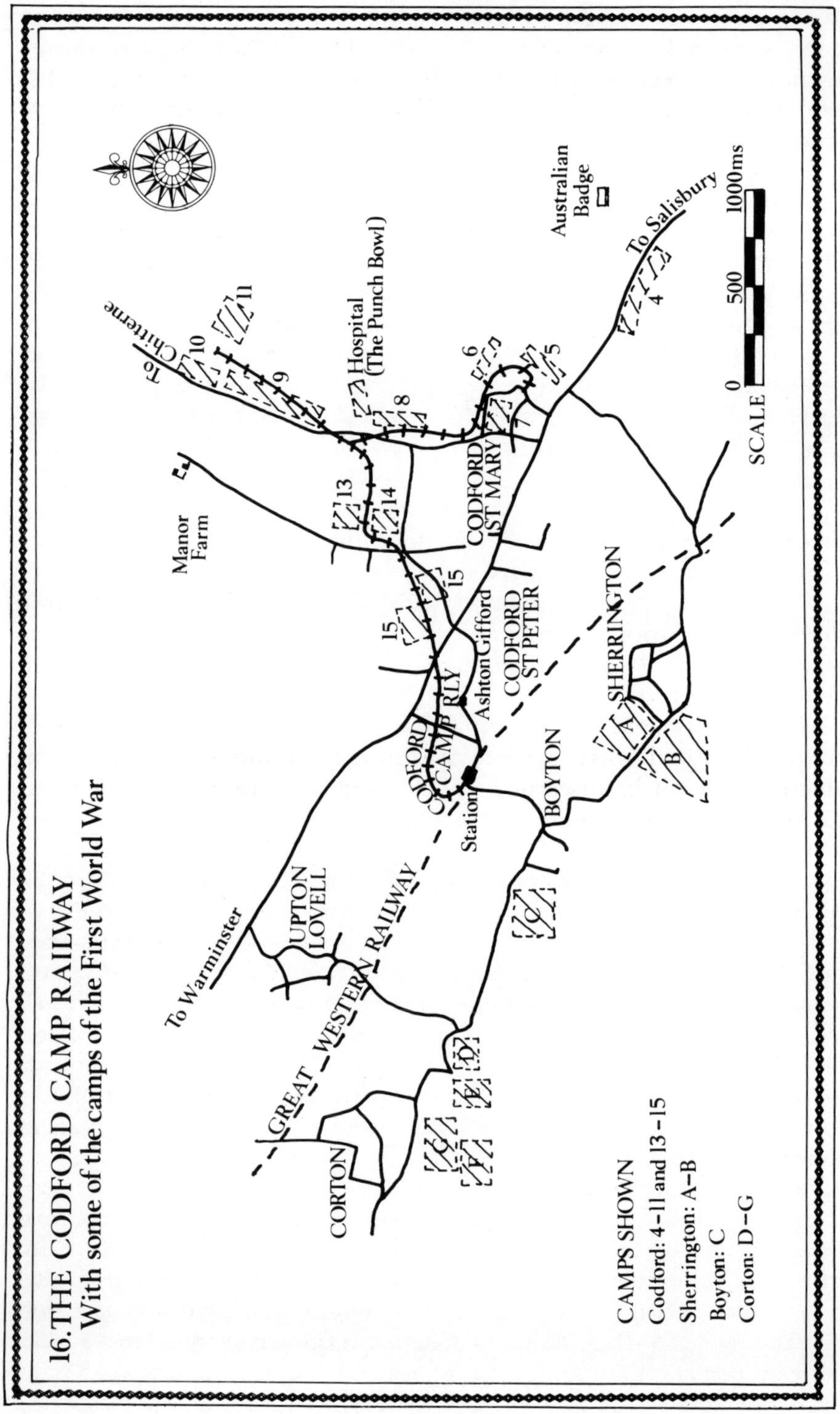

16. THE CODFORD CAMP RAILWAY
With some of the camps of the First World War

station buildings at Bulford were demolished and an accommodation block for the Salisbury Plain Area Works Office of the Property Services Agency, was erected on the site. This was opened on 18th September 1969 and to mark its connection with the railway, an upper quadrant signal was placed at the entrance to the approach drive.

The longest railway on the Plain that was operated by the War Department, was the Larkhill Military Railway which followed a roundabout course from Ratfyn near Amesbury to Druid's Lodge, a mile and a half to the south of Long Barrow Cross Roads. It was built in the autumn of 1914 and the spring of 1915 by Sir John Jackson whose firm was also responsible for the construction of the hutted camps at Larkhill. The railway began at a point some 700yds (640m.) north of Amesbury station where it joined the Amesbury–Bulford line at Ratfyn and close to which two engine sheds, a passing loop and a siding were built. The track then passed over the River Avon on a low viaduct and after running through a set of sidings, crossed the Amesbury–Upavon road (A345) at Countess. here it curved to the north-west and after three-quarters of a mile reached the reversing triangle where locomotives could turn around and which was located next to what are now the senior officers quarters at Strangways. Additional sidings were provided at the south-west end of the triangle while the line turned due north over Fargo Road, immediately to the west of what is now Lawson Road.

At the north end of Lawson Road and close to its junction with Colquhoun Road, was a 'station' consisting of a building and a platform which could still be discerned under a mass of brambles in 1986. At this point the main line forked to the west while a branch continued northwards as far as the Packway where it ended. After proceeding in a westerly direction for a quarter of a mile or so, the railway curved to the north, crossed the Packway, close to what is now Lawrence Road, and after completing a left-hand half circle rejoined it near the entrance to Roberts Barracks. From here the line ran besides the Packway until it reached what is now Ross Road where it branched away to the north and then due west to the Hamilton Battery. Some 700yds (640m.) beyond the Battery the main line turned due south while a branch continued to Rollestone Camp and shortly before reaching it a subsidiary branch turned south to the ammunition compound which at that time was on the north side of the Packway.

On crossing the Packway, the line followed a straight course to Fargo Military Hospital and after cutting through Fargo Plantation, ran across the Amesbury–Tilshead road (A360) some 900yds (820m.) to the east of Airman's Cross. On the south side of the road and parallel to it, a spur, about 450yds (410m.) in length, provided a connection with the Handley-Page hangars. The main track continued in a south-westerly direction to the group of tumuli on the northside of Winterbourne Stoke Clump, a small wood adjoining the Amesbury–Wyley

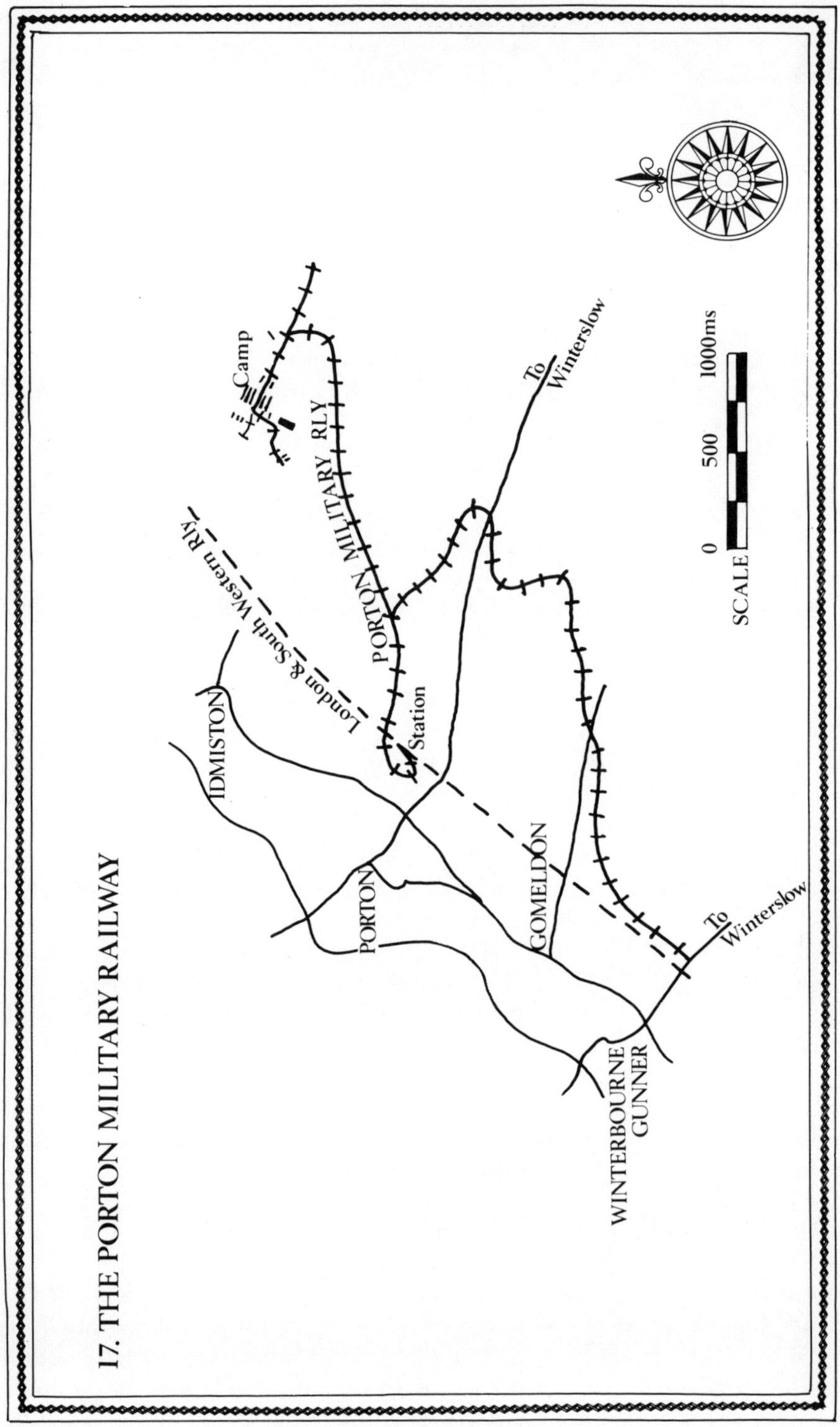

17. THE PORTON MILITARY RAILWAY

(*Flt.-Lt. J. O'Donnell*)

70. An aerial photograph of Upavon aerodrome, looking westwards, which was probably taken in 1913.

(*Imperial War Museum*)

71. Their Majesties King George VI and Queen Elizabeth visiting 1st Airborne Division at Netheravon on 21st May 1942. From left to right: HM The King; Maj.-Gen. F. R. M. Browning GOC 1st Airborne Division; Gen. Sir Bernard Paget, C-in-C Home Forces; HM The Queen; a Paratrooper; Brig. E. W. C. Flavell, Cmdr. 1st Parachute Bde.

(*Lt.-Col. D. J. Wilson*)

72. Some of the original 'black and white' buildings in Lower Airfield Camp, Netheravon.

(*Lt.-Col. D. J. Wilson*)

73. The carved 'wings' of the Royal Flying Corps over the entrance to the Officers' Mess, Netheravon.

(*H. C. Casserley*)

74. A train headed by ex-Great Western Railway 2–6–0 No. 6320, about to leave Tidworth Station on 1st September 1952.

(*J. Moloney*)

75. Tidworth Station facing Ludgershall. On the right is the military platform used for troop movements.

(*H. C. Casserley*)

76. Amesbury Station on 23rd March 1963. The two centre lines lead to Newton Tony.

(*H. C. Casserley*)

77. The view from Bulford Station looking towards Amesbury. On the left is an Adams 0–4–4 tank locomotive No. E.69 of the Southern Railway, 28th April 1928.

(*H. C. Casserley*)

78. The platform at Bulford Camp which was situated at the end of Horne Road. Photographed on 14th May 1955 when a train was run for railway enthusiasts.

(*N. D. G. James*)

79. The only evidence of Bulford Station in 1986. A railway signal erected at the entrance of the approach road to the offices of the Property Services Agency, built on the station site in 1969.

(*Imperial War Museum*)

80. Loading a 6 inch howitzer onto a railway wagon in 1939.

(*T. L. Fuller & Sons*)

81. The locomotive *Betty*, a saddle tank 0–6–0, which worked on the Tidworth Military Railway.

(Maj. M. W. Knight)

82. A view of the Royal Army Ordnance Depot at Tidworth in about 1930. At the top of the photograph can be seen the line to Ludgershall and sidings on which wagons are standing. In the bottom right hand corner part of the line to the barracks can be seen.

(*G. Ward*)

83. The Sutton Veny Camp Railway. The narrow gauge track which was used for distributing materials from unloading centres can be seen on the left.

(*P. Cole*)

84. The Codford Camp Railway.

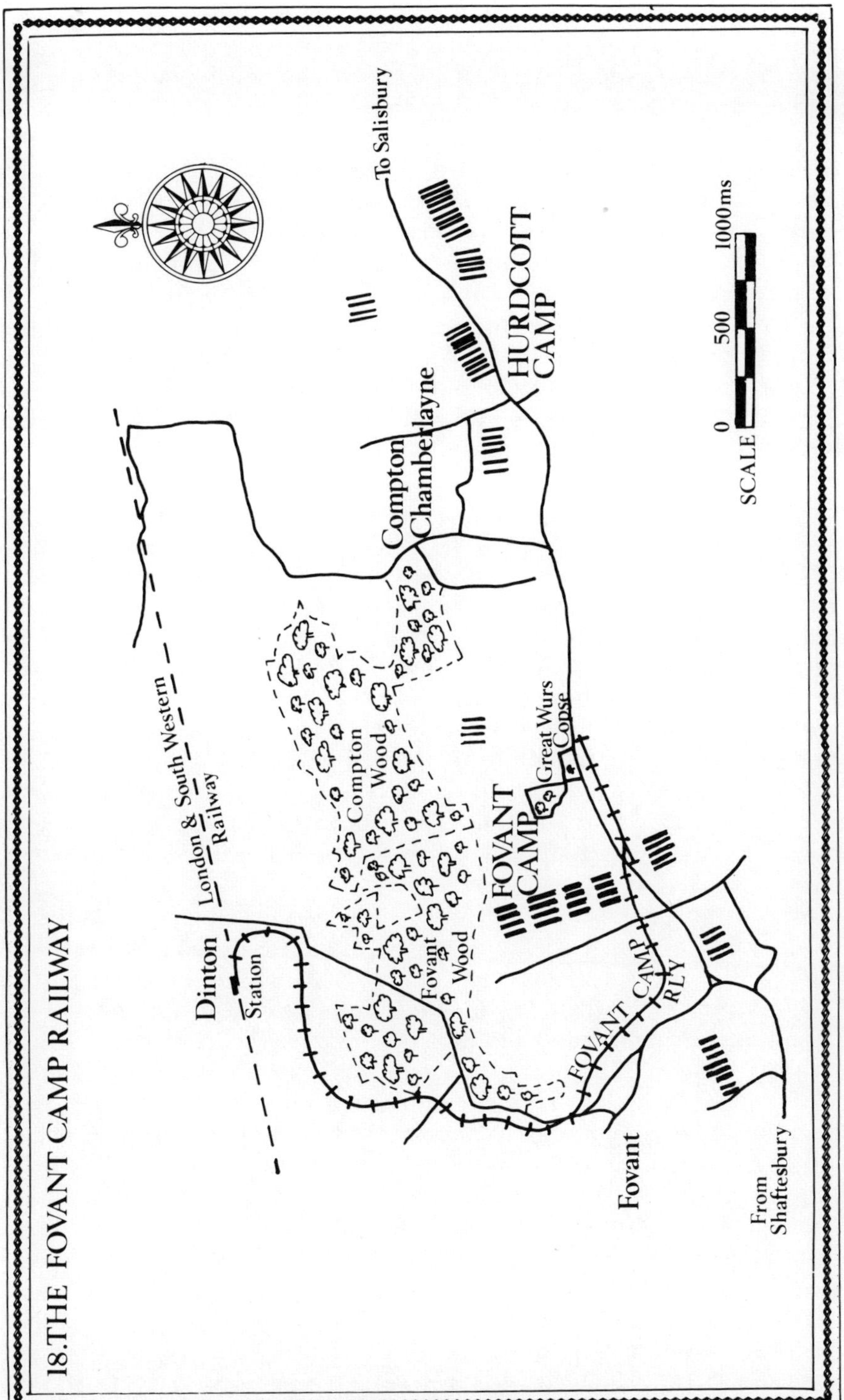

18. THE FOVANT CAMP RAILWAY

road (A303). However, nearly half a mile before reaching this road, a branch curved away to Stonehenge aerodrome. Beyond the Amesbury road the railway ran along the east side of the Salisbury road (A360) until it terminated about 600yds (550m.) north of Druid's Lodge where there was a hutted camp on the west side of the road. At this point the line provided a link with Lake Down Aerodrome. The length of what has been referred to as the 'main line' which ran from Ratfyn to Druid's Lodge, was approximately 7.31 miles but if the various branches and the two additional sides of the reversing triangle, are included (but not the sidings and loop lines) the distance is increased to 10.78 miles.

There were two locomotives which apparently belonged to the company operated by Sir John Jackson since they were first used during the construction of the railway. These were a 0–4–0 saddle tank named *Queen Mary* and a 0–6–0 side tank which was named *Salisbury*. After the line was completed, *Queen Mary* continued to work on the railway and it is probable that *Salisbury* was also retained for at least a year or so, if not permanently. The line was operated by a railway company of the Royal Engineers who occupied a camp adjoining the level crossing over the Amesbury road at Countess. The section to the south of Fargo Hospital which included the Stonehenge branch and the line to Druid's Lodge, was the first to be closed and the Ordnance map shows that the track had been lifted by 1923.[41] The remainder of the line had ceased to operate by 1928 and it would appear that most of the track had been removed by 1932. However it was still possible to see some evidence of the railway in 1986, mainly in the form of embankments and cuttings between Countess Crossing and the Packway and including most of the reversing triangle and the platform in Lawson Road. A longer and more detailed description of this railway will be found in *Gunners at Larkhill*.[42]

The Sutton Veny Camp Railway covered a route distance of approximately $3\frac{3}{4}$ miles and work on laying the track had probably begun by the end of 1914, since the construction of a similar camp railway at Codford was well advanced by mid-October.[43] Until it was closed, Heytesbury Station stood besides the bridge which carried the road from the village to Tytherton, over the railway. From this station the line ran close to the south bank of the Wylye, crossing the Sutton Veny–Heytesbury road (B3095) at Norton Waters (grid reference 909427) and continued in a westerly direction past Cooper's Bottom to the Bishopstrow road at reference 893419. It then continued through Eastleigh Wood crossing Five Ash Lane about 200yds (183m.) north-west of Raester's Farm and swung south and south-west through Southleigh Wood, terminating at what is now Southleigh Farm.[44]

The hospital which did not open until 1916, occupied the northern end of the Common between the point at which the railway crossed the Bishopstrow road and Marsh Road. Arrangements were made for ambulance trains to run alongside the reception centre so that the wounded could be transferred to the wards with the minimum delay.[45] To the south of the hospital was the main railhead at which most of the wagons were unloaded and from where stores, rations and so on were distributed over light 2ft. (610mm.) gauge track to different camps.

In August 1914 it was clear that improved railway facilities were needed at Codford station and by October, the construction of new sidings were reported to be 'well in hand'.[46] However the sidings were only a beginning and the following issue of the *Warminster Journal* reported that 'the extension of the railway is being pushed ahead rapidly',[47] and the line took the following route. On leaving Codford Station (grid reference 955400), the Codford Camp Railway swung eastwards to complete a half circle, passing to the north of Ashton Gifford before crossing the main road about 700yds (640m.) north-west of Codford St. Peter Church. From here it continued westwards through Camp 15, to cross the road from Codford St. Peter to Manor Farm at reference 966404. The line then turned north for some 200yds (180m.) before bearing due east between Camps 13 and 14 and over the Chitterne Brook to reference 971405. At this point it swung north again over the Codford-Chitterne road at 972408 to the north-west of Codford Hill where this section of the line terminated at reference 977413. At grid reference 972406 the railway divided and a branch ran southwards over the Chitterne road, opposite the Punch Bowl, and then besides the road to 972400. Here it bore to the east along the present road on the north of Codford St. Mary, running between Camps 6 and 7 to reference 976399 where it swung south before it ended besides Camp 5.[48] The total route length was a little over $2\frac{1}{2}$ miles.

In their Report of 1st June 1918, the Lands and Buildings Reconstruction Committee recorded, in a section entitled 'Light Railways and Sidings' an entry in respect of a 'light railway to Boscombe Down Aerodrome'.[49] Although No.6 Training Depot Station occupied Boscombe Down airfield on 1st October 1917, the aerodrome buildings were not completed until November 1918 and in view of the date of this Report, the railway was probably built at the end of 1917 or early in 1918. The Ordnance map of *c.*1918 shows the railway running in a south-westerly direction from a point on the Amesbury–Newton Tony line, 1.25 miles south-east of Amesbury station.[50] The line was 530yds (485m.) long and the Ordnance sheet of 1926 shows that a siding of some 300yds (275m.) was added later although both the line and siding had been removed by 1923.[51]

Although the Porton Military Railway and the Fovant Camp Railway lay beyond the boundaries of Salisbury Plain, as defined in Chapter 1, they have been included in order to complete the record of War Department railways on or near the Plain. The railway at Porton began at its junction with the main London & South Western line, some 200yds (180m.) north-east of Porton station. After proceeding for about 250yds in the direction of Salisbury, it turned in a semi-circle to cross the main line by a bridge at grid reference 195365. Beyond the bridge the track continued eastwards for about 680yds (620m.) to a junction where one line branched to the south, while the other continued in a north-easterly direction to the camp buildings which were situated in the area of 213371. The southern branch of the railway proceeded in the direction of Portondown Farm (reference 208360) but 200yds before reaching it, the line curved to the west and then south to a point at approximately 204357. Here it turned west again and after crossing the Gomeldon road, half a mile to the east of the main line, swung south parallel to the railway and ended at the Winterbourne Gunner–Winterslow road 100yds east of the South Western Railway. The length of the line from Porton station to the camp buildings was approximately 2.35 miles while the branch to Winterbourne Gunner from its junction with the camp line was about 2.09 miles, giving a total route distance of 4.44 miles. There was, in addition, a siding 420yds (384m.) long to the east of the camp.

Fovant Camp lay between the main Salisbury–Exeter line of the London & South Western Railway and the Salisbury–Shaftesbury road (A30) although some of the huts were sited just to the south of the road. The Ordnance map[52] shows that the main camp area was situated between the villages of Fovant and Compton Chamberlayne, to the east of which was Hurdcott Camp. The Fovant Camp Railway connected with the main line at Dinton station, eight and a half miles west of Salisbury, and after leaving the station, it curved to the south over the River Nadder and then west besides the river, for about 800yds (730m.). At this point the line turned due south beside Fovant Wood, and opposite Fovant village it branched sharply to the east and crossed the Shaftesbury road (A30) about 900yds (820m.) east of its junction with the Fovant–Dinton road. It then ran parallel to the main road for about 750yds terminating just beyond Great Wurs Copse at refernce 020291. To the south of the main road was a siding about 115yds long which lead to what was apparently an engine shed. The total length of the line, excluding this siding, was three miles and as the camp was built in 1915,[53] the railway was presumably constructed at the same time. The track was lifted after the First War when the camp was dismantled.

The military railways which have already been described in this chapter are now defunct and their tracks lifted but in 1986 there was still one line running which was operated by one of the services. This was a 2ft. gauge railway which connected a siding and transfer point near Dinton with the Royal Air Force Storage Depot at Chilmark. Altogether there are over nine miles of narrow gauge and two miles of standard gauge track but details of the layout are not available. However, an article by Flt.-Lieut. A.F. Saunders RAF, describing the line and the rolling-stock, was published in *The Railway Magazine* in 1976[54]

CHAPTER FOURTEEN

Salisbury Plain Today

Many people must have received a profound shock when they read the following report which appeared in a prominent position on the front page of *The Wiltshire Gazette*, on 10th April 1930.

> We are informed on excellent, though not official, authority that the Government intend selling off the whole of the land belonging to the War Office on and around Salisbury Plain. The news may seem almost incredible, as for some time past there have been statements circulating of an opposite character – that the War Office property on the Plain will be further extended in the direction of Imber. We do not, however, publish the information we do now, without good reason to believe in its correctness.
>
> This reversal of a policy that has been in operation for the last thirty years, and was regarded as practically irrevocable, is amazing. It will have far-reaching effects, the nature of which cannot at present be estimated, on the whole of the large area concerned. Our information further is, that much other War Office property in the country will be sold. But as to that we have no details.
>
> Salisbury Plain is now largely used for heavy gun practice, and the question arises, where will this take place if the Plain is given up? We understand that there is sufficient Government land in Yorkshire. The Plain is also used very largely for the training of the mechanized units and for summer camps of the Territorial Army, as well as playing an important part in the more elaborate exercises. It is further understood that a number of the Regimental Depots will be closed. This has been 'in the air' for a considerable time, and it is understood to be now decided upon.[1]

The day after this report was published, the *Gazette* received a categorical denial from the War Office and its next issue contained this account.

> With reference to the statement published in last week's *Wiltshire Gazette* that the Government intended to sell the War Office property on and around

Salisbury Plain, the following telegram was received on Friday, addressed to the Editor:–

'There is not the slightest truth in the statement in the *Wiltshire Gazette* of April 10th that all War Office property on Salisbury Plain is to be sold. – H.J. Creedy, Permanent Under-Secretary of State for War'.

In the face of this unqualified contradiction we do not persist in our statement. Nevertheless we say emphatically that the matter has been under the consideration of the Government, whether the War Office knows it or not.

Our information was explicit, and came from a quarter that has early and accurate knowledge of all pending transactions in the way of the sale of large properties. The information was even so detailed as to include the fact that the Salisbury Plain property was to be sold 'in sections' and it is a coincidence that a syndicate has made an offer to purchase the land on Salisbury Plain but that no sale has been effected 'yet'.

The explanation of the matter probably is that the Chancellor of the Exchequer was looking round for money to help to balance his budget (the Plain property is a big asset) and that surveyors' reports were called for – for that there are documents concerning a possible sale we have positive evidence. In this connection it must be remembered that there is a strong anti-militarist element in the Government, and what would be unbelievable with a Government of another 'colour' is quite possible with the present one.

We may add that we do not regret having published the statement of last week. With the information before us we should not have been justified in withholding from our readers, many of whom are personally interested, knowledge of what appeared to be pending. We think they know the *Gazette* and its methods well enough not to believe that it would indulge in a 'stunt', and that what was done, was done with a full sense of responsibility.

One of the several 'contradictions' that appeared in our contemporaries stated that no further purchases will be made by the War Office in the Salisbury Plain area. That remains to be seen. But it is notorious that negotiations for further extensive acquisitions on the western side of the Plain have been proceeding for some time. They have lately been suspended, but whether the intention to purchase has been abandoned remains to be seen. As we said last week, the negotiations made the reported policy of selling seem the more remarkable.[2]

The only other reference to this matter appeared in the *Gazette* on 8th May when it was reported that the military correspondent of *The Morning Post* had confirmed that the purchase of land between Tilshead and Imber had not taken place. He added that 'a training area for the armed forces of the Crown is not regarded in these days as an essential national requirement'.[3] Some idea of the political background of this period can be formed from the fact that the Conservative Government under the leadership of Stanley Baldwin, had resigned in 1929 and had been replaced by a Labour Government under Ramsey Macdonald. In 1930 the occupation of Germany by allied troops had come to an end and France had evacuated the Rhineland and these developments were perhaps interpreted as evidence of a lasting peace in Europe. However those who took a more realistic view, noted two significant events which occurred during that year: in Germany, 107 members of the new Nazi party were elected to the Reichstag while the ratification of the London Naval Treaty resulted in the resignation of the Chief of the Japanese Naval Staff and the assassination of the Japanese Prime Minister. If the allegation that the Government intended to sell the Plain which was made by *The Wiltshire Gazette* was true, the fact that the sale did not materialize was probably due to the Government's resignation in 1931 when it was replaced by a National Government.

The extent to which the Army has used the Plain for training since the beginning of this century, has fluctuated considerably. From 1900 to 1913 its use increased, partly due to the South African War but also as its potentialities for training were realized. During the First World War it became one of the most important training centres in the United Kingdom but from 1920 until 1937 it reverted to a period of relaxed peacetime soldiering. By 1938 there were signs of renewed activity which increased steadily up to the outbreak of the Second World War when the demands of the Plain reached their maximum. Between 1947 and the end of the 1960s the intensity of training diminshed but since 1970 the position has changed substantially and training is now carried on throughout the year. There are three principal reasons for this: many overseas training areas such as Cyprus and Libya are no longer available; more realistic training methods have been adopted especially by the infantry while more importance is now attached to tactical training.

Since 1977, the use of the Plain for training purposes has been under the control and management of Headquarters Salisbury Plain Training Areas (HQ SPTA), and to facilitate this, the training area has been divided into three parts. The western area known as SPTA (West) extends from the School of Infantry to the Tilshead–West Lavington road and includes the Imber ranges. The second area which forms the centre block, covers Larkhill and West Down and stretches eastwards from the Shrewton–West Lavington road to within a short distance of the road from Amesbury to Upavon (A345). Beyond this road is SPTA (East) which extends to the eastern boundary of the land owned by the Ministry of Defence extending from Shipton Bellinger to Ludgershall and Everleigh.

If a unit or formation wishes to carry out training, it must apply to HQ SPTA for the use of an area for a specified period and a monthly conference is held in order to agree allocations. Since the Plain can be used by any unit in the United Kingdom, arrangements have to be made to accommodate them during their training and to cater for this, six training camps have been established at Knook, New Zealand Farm, Rollestone, West Down North, West Down South and West Down Depot respectively. The staff at Headquarters includes a Commandant, an Air Operations Officer, a Land Liaison Officer and a Grade 3 Staff Officer with three air traffic operations officers (communications to aircraft), a training area superintendent and a range superintendent. There are also Range Liaison Officers at the Royal School of Artillery at Larkhill and the School of Infantry at Warminster.

The Commandant is responsible for conservation over the whole of the training area. To fulfil the Ministry of Defence's responsibilities for conservation, to which much importance is attached, three conservation groups have been formed, namely, SPTA (West), Larkhill/West Down and SPTA (East), and all these groups have civilian members as well as personnel from units on and close to the Plain. These groups are organized into a varying number of disciplines or sub-groups which are for example concerned with special aspects such as archeology, botany, vertebrates, invertebrates, mammals, ornithology and so on. It is interesting to note that on those areas of the Plain where modern farming operations are restricted and to which access by the public is limited, wild flora and fauna flourish exceptionally well.[4]

The steps which were taken to create the War Department's estate on Salisbury Plain at the beginning of this century, have been described in Chapter 3 and it now covers some 91,000 acres (36,825 hectares) which includes 51 farms and over 5,500 acres (2,226 hectares) of woodland. The management of this very large estate is now the responsibility of the Property Services Agency of the Department of the Environment through the Defence Land Agent who is the local representative. His work which covers all the many aspects of estate management, also includes taking the necessary steps to ensure that the requirements of military training are met in so far as the use of land is concerned. He is also responsible for dealing with the farm tenants and with all matters relating to farming, houses, cottages and buildings. The woodlands are under the direction of the Chief Forester who is a member of the Land Agent's staff, and for management purposes these are divided into three categories – woodlands used for military training, normal estate woodlands and amenity woodlands. Since 1955, over five million trees have been planted on the Plain and in the garrison areas.[5]

After 1959 a number of the old camps which were no longer required, were demolished and these included Parkhouse Camp near Tidworth, Fowler Barracks and Busigny Barracks at Perham Down and Carter Barracks at Bulford. At the same time a great deal of rebuilding and development took place at Tidworth (1953), Larkhill (1960–1968), Warminster (1965–1969), and Perham Down (1974–1975) while a large number of married quarters were built in all the garrison areas. More recently plans have been made for modernizing and redeveloping the whole of Tidworth Garrison and work on Aliwal Barracks began in 1986. Development of a rather different kind was announced in 1984 when a proposal to build the replica of a German village on the Plain, was made public. Consisting of some ninety buildings, it is to be erected on Copehill Down about a mile south-west of Tilshead. The object is to provide realistic training for fighting in built up areas (FIBUA) and this 'village' will enable methods and drills to be used which cannot be practiced at Imber but it is unlikely to be in use before 1987.

During the past eighty years – a period which has included two world wars – thousands of troops have been stationed on Salisbury Plain and it is inevitable that some now lie in its cemeteries and churchyards. Many of the graves occur singly or in small groups but in some cases the numbers are many more. The Tidworth Military Cemetery which is the largest, contains four hundred and twenty-four graves of the First World War, one hundred and forty-seven of the Second War and one of the Falklands War, as well as a number of those who have died in peacetime. At Durrington Cemetery there are two hundred and twenty-six graves while in Sutton Veny and Codford St. Mary churchyards, one hundred and seventy-four Australians and sixty-six New Zealanders are buried. Except for one member of the Royal Naval Air Service and one soldier, all the thirty-five graves at Upavon are those of members of the Royal Air Force or Commonwealth Air Forces while among the graves in Netheravon churchyard are eight members of *HMS Daedalus* and *HMS Pembroke*.[6]

* * *

It is now ninety years since the War Office purchased the first land on Salisbury Plain and over the years the Army has gained a tremendous amount from the training facilities which have been provided. It can be said that the Plain occupies a unique position in the life of the Army and there can be few soldiers who have not trained on it at one time or another. No one can foresee the future but whatever it may hold, it is more than likely that soldiering on the Plain will continue for many years to come.

References

CHAPTER 1 THE GREAT PLAIN

1. Ella Noyes, *Salisbury Plain*, 1913, p.5.
2. N.D.G. James, *A History of English Forestry*, 1981, pp.95–96 and 109–110.
3. John Speed, *Map of Wiltshire*, 1610.
4. J. Andrews and A. Dury, *Map of Wiltshire*, 1st edn. 1773; 2nd edn. 1810.
5. T. Davis, *General View of the Agriculture of the County of Wilts*, 1794, frontispiece.
6. J. Cary, 'A New Map of Wiltshire, 1801' – *New English Atlas*, 1809.
7. *Kelly's Directory of Wiltshire*, 1899, p.2b.
8. Defence Land Agent, Durrington, by communication.
9. A.G. Bradley, *Round About Wiltshire*, 1st edn. 1907, pp.262–263.
10. E.V. Sargent, Record of Enford Farm, 1896–1978, unpublished MS.
11. Davis, *Agriculture of Wilts*, p.22.
12. J.M. Wilson (editor), *The Rural Cyclopedia*, vol.VI, 1849, p.188b.
13. Sargent, Enford Farm.
14. D. Defoe, *Tour through the whole island of Great Britain*, 1724–1727, Letter III, p.26.
15. Davis, *Agriculture of Wilts*, p.41.
16. John Aubrey, *The Natural History of Wiltshire*, edited by John Britton, 1847, p.107.
17. Ibid., p.54.
18. William Cobbett, *Rural Rides in the Counties of ... Wilts ...* A new edition with notes by James Paul Cobbett, 1853, p.31.

CHAPTER 2 THE NEW ORDER

1. Lieut.-Col. H.N. Cole, *The Story of Aldershot*, 1951, p.26.
2. Ibid., pp.28–29.
3. Ibid., p.31.
4. Ibid., p.82.
5. Maj. F. Myatt, *History of the Small Arms School Corps*, undated, pp.8, 10, 12.
6. Admiral of the Fleet, Lord Fisher, *Records*, 1919, p.19.
7. Maj.-Gen. Sir Charles Callwell and Maj.-Gen. Sir John Headlam, *The History of the Royal Artillery*, vol.I, 1931, p.229.
8. Ibid., p.233.
9. Ibid.
10. Byron Farwell, *Queen Victoria's Little Wars*, 1974, pp.364–371.
11. *Reserve Forces Act*, 1882. 45 & 46 Vict., c.48.

12. *The Military Manoeuvres Act*, 1897. 60 & 61 Vict., c.43.
13. *The Military Manoeuvres Act*, 1911. 1 & 2 Geo.5., c.44.
14. *The Manoeuvres Act*, 1958, 7 & 8 Eliz.2., c.7.
15. 'The Official Report on the Salisbury Manoeuvres in 1898' *Journal of the Royal United Services Institute for Defence Studies*, vol.XLIII, 1899, p.293.
16. Benjamin Sloper (Edward Slow), *Tha Military Manoovers in tha Nayberhood a Zalsbury*, 1907.
17. 'The Autumn Manoeuvres', *The Salisbury and Winchester Journal*, 14th Sept. 1872.
18. Ibid.
19. 'Grand Review of Boscombe Down', *Salisbury Journal*, 10th Sep. 1898.
20. Anonymous, *Mrs. Brown at the Review being a Rhymed Account of the March Past on Boscombe Down, September 8th 1898.*
21. Col. G.F.R. Henderson, *The Science of War – A collection of Essays and Lectures, 1891–1903*, edited by Col. N. Malcolm, 1933, p.396.
22. Ibid., p.409.
23. Ibid., p.396.
24. *The Military Lands Act*, 1892. 55 & 56 Vict. c.43.
25. 'Military Training Ground near Salisbury', *Salisbury Journal*, 30th Jan. 1897.
26. Ibid.
27. 'The Military Training Ground on Salisbury Plain', *Salisbury Journal*, 6th Feb. 1897.
28. Ibid.
29. 'Extracts from Secretary of State's Minutes'. *Minutes of Proceedings of the War Office Salisbury Plain Committee*, 1897–1902.
30. Minutes of First Meeting, *Salisbury Plain Committee*, p.6.
31. 'Extracts from Secretary of State's Minutes', *Salisbury Plain Committee.*
32. 'The Military Training Ground on Salisbury Plain', *Salisbury Journal*, 6th Feb. 1897.
33. 'The Military Training Ground on Salisbury Plain', *Salisbury Journal*, 13th Feb. 1897.
34. Ibid.
35. Ibid.
36. *Minutes of Salisbury Plain Committee*, Appendix No. 6 to minute 419, pp.116–119.
37. Ibid.
38. 'Military Wiltshire – Map of the War Office Land', *The Devizes and Wiltshire Gazette*, 29th June, 1899.
39. *Minutes of Salisbury Plain Committee*, Appendix No.8 to minute 503, p.151.
40. Defence Land Agent, Durrington, by communication.
41. Ibid.
42. Ibid.

CHAPTER 3 LAYING THE FOUNDATIONS

1. *Minutes of Proceedings of the War Office Salisbury Plain Committee*, 1897–1902.
2. Ibid., p.8, minute 7.
3. Ibid., p.142, minute 498.
4. Ibid., p.8, minute 8.
5. Ibid., p.23, minute 43.
6. Ibid., p.24, minute 48.
7. Ibid., p.52, minute 160.
8. Ibid., p.85, minute 322.
9. Ibid., p.89, minute 339.
10. Ibid., p.141, minute 492.
11. Ibid., p.111, minute 414.
12. Ibid., p.13, minute 16.

13. Ibid., p.17, minute 25.
14. Ibid., p.68, minute 226.
15. Ibid., p.41, minute 116.
16. Ibid., p.59, minute 194.
17. Ibid., p.88, minute 337.
18. Ibid., p.47, minute 135.
19. Ibid., p.56, minute 175.
20. Ibid., p.110, minute 412.
21. Ibid., p.87, minute 335.
22. Ibid., p.38, minute 97.
23. Ibid., p.50, minute 151.
24. Ibid., p.15, minute 24.
25. Ibid., p.29, minute 65.
26. Ibid., p.64, minute 215.
27. Ibid., p.46, minute 130.
28. Ibid., p.57, minute 182.
29. Ibid., p.46, minute 131.
30. Ibid., p.38, minute 99.
31. Ibid., p.113, minute 420.
32. Ibid., p.68, minute 228.
33. Ibid., p.140, minute 489.
34. Ibid., Appendix No.7 to minute 498, p.143.
35. Ibid., p.144, minute 502.
36. Defence Land Agent, Durrington by communication.
37. *The Parliamentary Debates*, Fourth Series, 60 VICTORIAE, vol.XLVII, 1897, p.1147.
38. *Minutes of Salisbury Plain Committee*, p.19, minute 27.
39. *Army Estimates for 1899–1900*, 17th February 1899, p.5.
40. 'Salisbury Plain Barracks', *The Devizes and Wiltshire Gazette*, 22nd June 1899.
41. 'Salisbury Plain Camps', *The Devizes and Wiltshire Gazette*, 20th July 1899.

CHAPTER 4 TIDWORTH – THE EARLY DAYS

1. Francis L. Bickley, *Victoria County History of Hampshire and the Isle of Wight*, vol.IV, 1911, p.391.
2. John Speed, *Map of Wiltshire*, 1610.
3. J. Andrews and A. Dury, *Map of Wiltshire*, 1st edn. 1773; 2nd edn. 1810.
4. Ordnance Survey Map to scale of one inch to one statute mile. Published 14th August 1817 by Col. Mudge at the Tower.
5. Ordnance Survey Map to scale of 1:10,560. Sheet XLVIII, 1877.
6. '50 Years Ago', *The Andover Advertiser and Northern Hants Gazette*, 27th August 1982.
7. Sir John Eardley-Wilmot, Bt., *A Famous Fox-Hunter, Reminiscences of the late Thomas Assheton Smith, Esq.*, 5th edn. 1893, p.7.
8. Ibid., pp.23–25.
9. Ibid., p.33.
10. Ibid., p.60.
11. Maj. J.F.L. Clarke, lately Honorary Secretary of the Tedworth Hunt, by communication, 27th August 1984.
12. Eardley-Wilmot, *A Famous Foxhunter*, pp.36–37.
13. Ibid., p.43.
14. Ibid., p.66.
15. Ibid.
16. Ibid., p.68.
17. Ibid.
18. A.G. Bradley, *Round About Wiltshire*, 7th edn. 1945, pp.224–225.

19. Eardley-Wilmot, *A Famous Foxhunter*, p.190.
20. Edith Buxton, *Reluctant Missionary*, 1968, pp.17–18.
21. Ibid., pp.23–24.
22. Ibid., p.25.
23. Ibid., p.26.
24. Conveyance of the Tedworth Estate from Sir John William Kelk, Bt., to the Secretary of State for the War Department, 29th September 1897, p.4, para.3.
25. Frederic Boase, *Modern English Biography*, 1965, vol.II–I to Q, pp.177–178.
26. 'The Military Development of Wiltshire – Plan and Particulars of the Tidworth Barracks', *The Devizes and Wiltshire Gazette*, 27th December 1900.
27. Ibid.
28. Ibid.
29. Ordnance Survey Map: scale two inches to one mile, *War Department Land on Salisbury Plain*, 1903.
30. *The Devizes and Wiltshire Gazette*, 27th December 1900.
31. 'On Salisbury Plain', *Andover Advertiser and Northern Hants Gazette*, 25th September 1903.
32. D. Bartholomew, *Midland and South Western Junction Railway*, vol.I, 1982, p.217.
33. *Devizes Gazette*, 27th December 1900.
34. Maj. E. Clarke-Smith, *The Larkhill Guide*, 1969, p.19.
35. *The Barrack Book 1929*, 118 General 714, The War Office, 1929, pp.82–85.
36. Ibid.
37. Clarke-Smith, *Larkhill Guide*, p.19.
38. *Andover Advertiser*, 25th September 1903.
39. The Hon. J.W. Fortescue, *A History of the British Army*, vol.XII, 1927, p.376.
40. Ibid., p.377.
41. Ibid., p.381.
42. Ibid., vol.V, 1910, pp.23–24.
43. Ibid., pp.33–34.
44. Ibid., p.32.
45. Ibid., vol.V, 1910, p.109.
46. Ibid., pp.124–126.
47. Ibid., p.128.
48. Ibid., vol.XI, pp.353–354.
49. Ibid., pp.357–363.
50. Ibid., p.366.
51. Ibid., vol.XII, p.256.
52. Ibid., p.272.
53. Maj.-Gen. Sir John Headlam, *The History of the Royal Artillery*, vol.III, 1940, pp.54–56.
54. Fortescue, *History of British Army*, vol.XIII, 1930, pp.266–268.
55. Ibid., pp.296–297.
56. Ibid., p.307.
57. Ibid., pp.311–312.
58. Ibid., pp.316–317.
59. Ibid., p.321.
60. Ibid., pp.325–326.
61. Ibid., pp.338–339.
62. Ibid., pp.341–342.
63. Ibid., vol.XII, 1927, pp.424–425.
64. Ibid., pp.425–426.
65. Ibid., p.446.

CHAPTER 5 TIDWORTH – DEVELOPMENT AND EXPANSIONS

1. *The Barrack Book 1929*, 118 General 714, The War Office 1929, p.82.
2. *The Army Medical Services Magazine*, April 1905.
3. *The Monthly Army List*, January 1904 and January 1905.
4. *Journal of the Royal Army Medical Corps*, Officer Distribution List, July 1907.
5. The Nursing Board, Queen Alexandra's Imperial Military Nursing Service, *Proceedings & Reports*, vol.7, April 1908 to March 1909, Nursing Inspection No.59.
6. Ibid.
7. Ibid.
8. Ibid.
9. Ordnance Survey Map, scale 1:10,560, Sheet SU24 NW, 1961.
10. Ordnance Survey Map, scale 1:10,000 Sheet SU24 NW, 1976.
11. *Barrack Book*, pp.82–83.
12. Maj. P.L. Rutherford QARANC by communication.
13. *Barrack Book*, p.84.
14. Letter to Assistant Principal Secretary's Department, the War Office from the Rev. W.S. Jaffray SCF (P) 7th August 1908.
15. Ordnance Survey Map: scale two inches to one mile, *War Department Land on Salisbury Plain*, 1903, revised 1894–1899.
16. Ordnance Survey Map, GSGS 2748, scale 1:20,000 published *c.*1918 and re-issued as GSGS 3906, scale 1:25,000 1931, sheet 44/16 SE Mapping and Charting Establishment RE, by communication October 1984.
17. Ordnance Survey Map, scale 1:10,560 Hampshire sheet XIV SE/Wiltshire sheet XLVIII SE; 1926 edition, revised 1923.
18. *Barrack Book 1929*, p.84.
19. Ibid.
20. Ibid.
21. *War Department Electricity System, Salisbury Plain Area: History, Development and Operation*, 1956, pp.1–2.
22. *Minutes of Proceedings of the War Office Salisbury Plain Committee*, 1897–1902, p.89, minute 339.
23. *The Monthly Army List*, June 1899, section (14), footnote.
24. *The Monthly Army List*, October 1900, section (15).
25. I.B.M. Hamilton, *The Happy Warrior – A Life of Gen. Sir Ian Hamilton, GCB, GCMG, DSO*, 1966, pp.233 and 239.
26. *Barrack Book 1929*. p.82.
27. Nikolaus Pevsner, *The Buildings of England – Wiltshire*, revised by Bridget Cherry, 1975, p.359.
28. Nikolaus Pevsner and David Lloyd, *The Buildings of England – Hampshire and The Isle of Wight*, 1967, pp. 602–603.
29. Order of Her Majesty in Council, 14th August 1972.
30. Letter to Assistant Principal Secretary's Department, the War Office from the Rev. W.S. Jaffray SCF (P) 14th July 1908.
31. Ibid.
32. Letter from the War Office to the Rev. W.S. Jaffray, 16th July 1908.
33. Letter from the War Office to the Rev. W.S. Jaffray, 30th July 1908.
34. Letter from the Rev. W.S. Jaffray to Headquarters Southern Command, 20th August 1909.
35. Lieut.-Col. C.C.G. Ross, Secretary, Church of England Board for the Welfare of Her Majesty's Forces within the Diocese of Salisbury, by communication 13th September 1984.
36. Rev. F.F.S. Smithwick, CF, 'Notable Garrison Churches. 12. St. Michael's Garrison Church, Tidworth'. *Royal Army Chaplains' Department Quarterly Journal*, July 1925, p.94.

37. *Thirteenth Report of the Church of England Board for the Welfare of Soldiers*, 1910, pp.1–2.
38. Rev. F.F.S. Smithwick, *RAChD Quarterly Journal*, p.94.
39. *Fourteenth Report of the Church of England Board for the Welfare of Soldiers on Salisbury Plain*, 1911, p.2.
40. Rev. F.F.S. Smithwick, *RAChD Quarterly Journal*, p.96.
41. *Fifteenth Report of the Church of England Board for the Welfare of Soldiers on Salisbury Plain*, 1912, p.2.
42. Rev. J.J. Holliman, CF, 'St. Michael's Garrison Church Tidworth', *Journal of the Royal Army Chaplains' Department*, vol.28, July 1984, p.13.
43. N. Pevsner and D. Lloyd, *Hampshire and The Isle of Wight*, 1967, p.603.
44. *The Ian Hamilton Soldiers' Home, Tidworth Visitor's Book*, Entry for 18th October 1964.
45. *New Wesleyan Soldier's Home, Tidworth Barracks, Salisbury Plain*, Appeal Brochure, June, 1907.
46. Ibid.
47. *Ian Hamilton Soldiers' Home – Vistor's Book.*
48. *Thirteenth Report Church of England Board*, 1910, pp.2–3.
49. Maj. G.W. Whittingham, National Secretary for Red Shield Services, by communication, 17th July 1984.
50. F. Stone, late Senior YMCA Representative Southern Command, *The Young Mens' Christian Association with H.M. Forces on Salisbury Plain*, unpublished MS, October 1984.
51. G.E. Wickens, National Council of YMCAs, by communication, 1st August 1984.
52. K. Pickernell, by communication.
53. 'Stars of Army Stage', *Southern Evening Echo*, 17th January 1984.
54. Annual Report of the Chief Fire Officer, Wiltshire County Council, 1963–1964, p.10.
55. Sir John Eardley-Wilmot, Bt. *A Famous Fox-Hunter, Reminiscences of the late Thomas Assheton Smith, Esq.*, 5th edn. 1893, p.68.
56. Col. A.C.T. White, VC. *The Story of Army Education 1643–1963*, 1963, p.33.
57. Ibid., pp.36–37.
58. Ibid., p.42.
59. Ibid., p.77.
60. *The Parliamentary Debates*, Fourth Series, 60 VICTORIAE, vol.XLVII, 1897, p.1147.
61. *Minutes of Proceedings of the War Office Salisbury Plain Committee*, 1897–1902, p.42, minute 199.
62. Ibid., p.61, minute 199.
63. Ordnance Survey Map: scale two inches to one mile, *War Department Land on Salisbury Plain*, 1903.

CHAPTER 6 TIDWORTH – WAR AND PEACE

1. *The Monthly Army List*, August 1914.
2. Gen. Sir Horace Smith-Dorrien, *Memories of Forty-Eight Years Service*, 1925, p.372.
3. S.C. Butler, 'Memories of Andover', *Test Valley and Border Anthology 14*, (Andover Local Archives Committee) 1979, p.325.
4. Maj.-Gen. Sir George Scott-Moncrieff, 'The Hutting Problem in the War', *The Royal Engineers' Journal*, vol.XXXVII, 1924, p.362.
5. Ibid., p.364.
6. Ibid., p.376.
7. A.F. Barnes, *The Story of the 2/5th Battalion, the Gloucestershire Regiment*, 1914–1919, 1920, p.24.
8. Ibid., pp.29–30.

9. Ibid., p.31.
10. Col. G.W.L. Nicholson, *Canadian Expeditionary Force 1914–1919*, Ottawa, 1964, p.39.
11. C.E.W. Bean, *Anzac to Amiens*, Canberra, 1948, p.203.
12. C.E.W. Bean, *The Australian Imperial Force in France 1916*, Sydney, 1929, p.169.
13. Ibid., p.168.
14. Ibid., p.169.
15. The Curator Official Records, Australian War Memorial Canberra, by communication, 19th December 1983.
16. Bean, *Australian Imperial Force*, pp.176, 168.
17. Ibid., p.170.
18. Bean, *Anzac to Amiens*, p.518.
19. Col. H. Stewart, *The New Zealand Division 1916–1919*, 1921, p.17.
20. Ibid., p.18.
21. *The Monthly Army List*, January 1921.
22. *The Monthly Army List*, January 1924.
23. *The Monthly Army List*, January 1932, p.45.
24. *The Monthly Army List*, January 1939, p.51.
25. *The Monthly Army List*, January 1942, p.64.
26. Lieut.-Col. C.R. Berkeley, *Southern Command Tidworth Tattoo*, pamphlet on its origin inception and subsequent development to 1930, unpublished MSS, 1932.
27. Ibid.
28. Ibid.
29. Ibid.
30. Ibid.
31. *Programme, Southern Command Tidworth Tattoo, 1935*.
32. Lieut.-Col. A. Hunt, by communication, 22nd November 1984.
33. 'Horse Trials and Tattoo at Tidworth', *The Andover Advertiser*, 20th May 1966.
34. 'George and Dragon at Tidworth Tattoo', *The Andover Advertiser*, 27th May 1966.
35. D.J. Miles, Estate Manager, Whitbread Wessex, by communication, 4th September 1984.
36. N.E. Trowell, Hon. Curator, NAAFI Historical Collection, by communication, 13th November 1984.
37. The General Secretary, The Royal British Legion, by communication, 21st December 1984.
38. Mrs. J. Keyes, by communication, 21st August 1984.
39. *The Andover Advertiser*, 12th September 1924, p.7.
40. Ordnance Survey Map, scale 1:10,560, Hampshire sheet XIV SE/Wiltshire sheet XLVIII SE; 1926 edition, revised 1923.
41. Ordnance Survey Map, scale 1:2,500, Hampshire sheet XIV–15/Wiltshire sheet XLVIII–15, 1936.
42. *Minutes of Proceedings of the War Office Salisbury Plain Committee*, 1897–1902, p.72, minute 254.
43. Ordnance Survey Map, GSGS 2748, scale 1:20,000 published *c.*1918 and reissued as GSGS 3906, scale 1:25,000 1931. Sheet 44/16 SE. Mapping and Charting Establishment RE, by communication.
44. Lieut.-Col. R.S.G. Perry, by communication, 1st March 1985.
45. Berkeley, *Southern Command Tidworth Tattoo.*
46. Brig. B.N.L. Fletcher, by communication, 28th January 1985.
47. Perry, by communication.
48. Ibid.
49. Maj. D.B.S. Brotherton, Garrison Adjutant Tidworth, by communication, 23rd January 1985.
50. 'Roundabout', *The Andover Advertiser*, 23rd November 1984.
51. Maj. E. Chapman, by communication, February 1984.

52. Ibid.
53. Programme for a Demonstration by the Experimental Armoured Force, Salisbury Plain, 12th July 1928.
54. Mrs. H.E.P. Colley, by communication, March 1984.
55. Ordnance Survey Map, 1:2,500, 1936.
56. Royal Warrant: Corps Warrant-Amendments, 22nd May 1939.
57. David Niven, *The Moon's a Balloon*, 1971, p.210.
58. Col. C.P. Stacey, *Six Years of War* – Official History of the Canadian Army in the Second World War, Ottawa, 1955, vol. I, p.419.
59. W.G. McClymont, *To Greece* – Official History of New Zealand in the Second World War 1939–1945, Wellington, 1959, pp.32–33.
60. Gavin Long, *To Benghazi* – Australia in the War of 1939–1945, Canberra, 1952, pp.305, 306, 307.
61. Ibid., p.308.
62. Ibid.
63. Ibid., p.307.
64. Ibid., p.308.
65. Ibid., p.310.
66. *Army and Air Force Installations outside Continental U.S.*, Department of the Army, 31st December 1947, pp.169–196.
67. Norman Longmate, *The G.I.'s – The Americans in Britain 1942–1945*, 1975, p.33.
68. *Order of Battle of the United States Army, World War II, European Theater of Operations: Divisions.* Paris, 1945.
69. Mrs. Theodore Roosevelt, Jr., *Day Before Yesterday*, 1959, p.428.
70. Ibid., p.432.
71. By communication, March 1984.
72. Longmate, *The G.I.'s*, pp.94–95.
73. 'Tidworth's Face is Lifted', *Soldier*, November 1953, vol.9, No.9, pp.8–9.
74. Maj. R.D. Baker, by communication, 26th January 1985.
75. The Principal Housing Officer (Management), Kennet District Council, by communication, 10th October 1984.

CHAPTER 7 PERHAM DOWN AND LUDGERSHALL

1. *Minutes of Proceedings of the War Office Salisbury Plain Committee*, 1897–1902, p.101, minute 380.
2. Ibid., p.61, minute 199.
3. Ordnance Survey Map: Scale two inches to one mile, *War Department Land on Salisbury Plain*, 1903.
4. A. Kennedy, *The Forces Postal History Society Newsletter 171*, Spring 1982.
5. Brig.-Gen. C. Ballard, *Smith-Dorrien*, 1931, p.134.
6. *The Barrack Book 1929*, 118 General 714, The War Office, 1929, p.82.
7. C.E.W. Bean, *The Australian Imperial Force in France 1916*, Sydney, 1929, pp.169–170.
8. *The Monthly Army List*, September 1921.
9. *The Monthly Army List*, July 1922.
10. *The Monthly Army List*, November 1922.
11. *A Short History of the Royal Tank Corps*, 5th edn. 1938, p.97.
12. 'The Corps Budget – Fifth Tank Battalion', *The Tank Corps Journal*, No.38, vol.4, June 1922, p.40.
13. Ibid.
14. *The Barrack Book 1921*, General Number 16/4501, p.74.
15. Ordnance Survey Map, scale 1:10,560, Hampshire Sheet XIV.SE/Wiltshire Sheet XLVIII, SE; 1926 edition revised 1923.
16. *Army Estimates* for 1921–1922 and 1924–1925.

17. *The Royal Tank Corps Journal*, vol.VII, January 1926, p.250.
18. *The Barrack Book 1929*, 118 General 714, The War Office 1929, p.82.
19. *Army Estimates* for 1928–1929.
20. *Barrack Book 1921*, p.74.
21. *Barrack Book 1929*, pp.82–83.
22. 'General Congreve reviews 3rd Division ... Battle at Perham', *The Andover Advertiser*, 23rd May 1924.
23. *Report on Training of 1st Brigade, Royal Tank Corps*, Part I, para.1, 1931.
24. *Short History Royal Tank Corps*, 1938, p.102.
25. Ibid., p.103.
26. Captain B.H. Liddell Hart, *The Tanks*, vol.I, 1959, p.317.
27. Mrs. M.B. Berkery and R.V. Peters, by communication.
28. L.E. Trevett, by communication.
29. *Army Estimates* for 1934–1935 and 1939–1940.
30. *Royal Tank Corps Journal*, vol.XVI, April 1935, p.347.
31. Property Services Agency, Tidworth, by communication.
32. *The Tank*, No.217, vol.19, May 1937, p.14.
33. *The Tank*, No.227, vol.19, March 1938, p.343.
34. *The Tank*, No.230, vol.20, June 1938, p.37.
35. *The Tank*, No.231, vol.20, July 1938, p.67.
36. Liddell Hart, *The Tanks*, vol.II, 1959, p.6.
37. Ordnance Survey Map, scale 1:2,500, Hampshire Sheet XIV.16/Wiltshire Sheet XLVIII.16, 1936.
38. Liddell Hart, *The Tanks*, vol.I, 1959, p.168.
39. Ordnance Survey Map, scale 1:10560, Hampshire Sheet XIV SE/Wiltshire Sheet XLVIII SE, 1926 edition, revised 1923.
40. Property Services Agency, Tidworth, by communication.
41. *The Tank*, No.240, vol.20, April 1939, pp.379 and 384.
42. *Paths of Armor*, History of the 5th Armored Division United States Army, 1950, pp.37–39.
43. Norman Longmate, *The G.I.'s – The Americans in Britain 1942–1945*, 1975, p.350.
44. Property Services Agency, Tidworth, by communication.
45. British Army News Service – Institute of Army Education, 7th May, 1968.
46. Property Services Agency, Tidworth, by communication.
47. R. Allen Brown, H.M. Colvin and A.J. Taylor, *The History of the King's Works*, vol.II, 1963, p.729.
48. 'Leland's Journey through Wiltshire AD 1540–42' with notes by the Rev. J.E. Jackson, *The Wiltshire Archaeological and Natural History Magazine*, vol.I, 1854, p.179.
49. John Speed, *Map of Wiltshire*, 1610.
50. T.B. Sands, *The Midland and South Western Junction Railway*, reprinted 1979, p.10.
51. Ibid., p.12.
52. *Minutes of Proceedings of the War Office Salisbury Plain Committee*, 1897–1902, Appendix 6 to minute 419, pp.117–118.
53. Ibid., p.97, minute 360.
54. A.G. Bradley, *Round about Wiltshire*, 1st edn. 1907, p.227.
55. Ordnance Survey Map: Scale 1:2,500, Wiltshire Sheet XLVIII–12, 1936.
56. *Guide to Vehicle Depot Ludgershall*, undated, p.1.
57. Ibid., pp.1, 4, 5.
58. Commandant, Defence Medical Equipment Depot, by communication, 27th November 1984.
59. Ibid.
60. *The David Bruce Laboratories, East Everleigh near Marlborough, Wilts*, pamphlet, 1970.
61. Ibid.

62. Ibid.

CHAPTER 8 BULFORD – THE CURTAIN RISES

1. Ordnance Survey Map: scale two inches to one mile, *War Department Land on Salisbury Plain*, 1903, revised 1894–1899.
2. Ordance Survey, Southampton, by communication, 21st October 1983.
3. Ordinance Survey Map: scale 1:50,000, Sheet 184, First Series 1979.
4. *Minutes of Proceedings of the War Office Salisbury Plain Committee*, 1897–1902, Appendix No.6 to minute 419, p.116 and Appendix to minute 508, p.149.
5. 'The War Office and Salisbury Plain. Encampment of Cavalry at Bulford', *The Salisbury and Winchester Journal*, 10th July 1897.
6. 'Camp at Bulford. Cavalry Manoeuvres' *The Salisbury Journal*, 14th August 1897.
7. Ibid.
8. Mary Hamilton, *The Silver Road*, 1951, pp.124–125.
9. Ordnance Survey Map, *War Department Land on Salisbury Plain*, 1903.
10. *The Barrack Book 1929*, 118 General 714, p.62.
11. *War Office Novel*, 1902–1903.
12. Ibid., 1909.
13. Col. F.M. Hill, by communication, 18th September 1984.
14. Hamilton, *The Silver Road*, p.129.
15. 'On Salisbury Plain', *Andover Advertiser*, 25th September 1903.
16. *Kelly's Directory for Wiltshire*, 1903, p.46.
17. *War Office Novel*, 1904–1905, p.129.
18. *Bulford Garrison Plumber's Work Books*, 1902–1904.
19. Ordnance Survey map, scale 1:2,500, Wiltshire Sheet LV.9, 1939.
20. E. Hall, by communication, 21st February 1984.
21. 'The Military on Salisbury Plain', *Salisbury Journal*, 9th May 1903.
22. *Abstract of the Recommendations of Commissions, Committees & Conferences relating to Army Affairs*, 1900–1920, p.13, *Mounted Infantry*.
23. 'On Salisbury Plain', *Andover Advertiser*, 8th May 1903.
24. 'On Salisbury Plain', *Andover Advertiser*, 25th September 1903.
25. *Minutes of Salisbury Plain Committee*, p.63, minute 204.
26. Ibid., p.45, minute 129.
27. D. Featherstone, *Weapons and Equipment of the Victorian Soldier*, 1978, p.24.
28. Ibid., p.25.
29. Ibid., p.30.
30. F. Myatt, *The British Infantry 1660–1945*, 1983, p.135.
31. *The Parliamentary Debates*, Fourth Series, 60 VICTORIAE, vol.XLV, 1897, p.495.
32.. Maj. M.J. Fogwell, by communication, 11th April 1984.
33. *Minutes of Salisbury Plain Committee*, p.39, minute 105.
34. Ibid., p.45, minute 129
35. Ibid., pp.51–52, minute 159.
36. Ibid., p.139, minute 484.
37. *Army Estimates* for 1909–1910.
38. Ordnance Survey Map, scale 1:10,560, Hampshire Sheets XXII.NW and XXII.SW and Wiltshire Sheets LV.NW and LV.SW, 1925 edn. revised 1923.
39. Ordnance Survey Map, scale 1:10,560, sheets SU24·NW and SU24 SW, 1961, revised 1957.
40. Plan No.A/61/70 prepared by Ministry of Public Buildings and Works, Area Office, Bulford 5th March 1970.
41. Ordnance Survey Map, scale 1:50,000 sheets SU14 SE, SU24 NW, Su24 SW, 1976 revised 1973–1975.
42. 'But he hit his mark', *Andover Advertiser*, 8th April 1910.
43. *Memorandum of Secretary of State relating to the Army Estimates for 1903–1904*, Cd 1416, 1903.

44. *Report on the Account of Army Expenditure for 1904–1905*, p.120, 1906.
45. A.G. Bradley, *Round about Wiltshire*, 1st edn. 1907, p.241.
46. *Army Estimates* for 1909–1910.
47. *Army Estimates* for 1910–1911.
48. *War Department Electricity System, Salisbury Plain Area: History, Development and Operation*, 1959, p.5.
49. Ordnance Survey Map: scale 1:2,500, Wiltshire Sheet LV.9, 1924.
50. *The Monthly Army List*, January 1904.
51. 'Salisbury Plain Notes', *Salisbury Journal*, 9 April 1904.
52. *The Monthly Army List*, April 1908.
53. *The Barrack Book, 1921*, p.60.
54. Ordnance Survey Map: scale 1:2,500, Wiltshire Sheet LV.9, 1939.
55. *The Barrack Book 1929*, p.62.
56. Archdeacon H.W. Carpenter, 'The Garrison Church, Bulford' *Quarterly Journal of the Royal Army Chaplains Department*, vol.5, no.41, January 1936, p.202.
57. Brig.-Gen. Sir James E. Edmonds, *History of the Great War – Military Operations France and Belgium 1915*, vol.2, 1928, p.367.
58. The Rev. W.R.F. Addison, 'Notable Garrison Churches, St. George's Garrison Church, Bulford', *The Journal of the Royal Army Chaplain's Department*, vol.3, no.11, July 1924, p.437.
59. Carpenter, *Quarterly Journal R.A. Ch.D*, vol.5, 1936, p.202.
60. Insert in 'Contents and Illustrations', *The Journal of the Royal Artillery*, vol.LV, No.1, April 1928.
61. Lieut.-Col. A.N. Prestige, 'Bulford Garrison Church', *Gunner*, No.38, January 1974, p.27.
62. The Rev. J. O'Hara, Senior RC Chaplain in charge of Records, Military Vicariate of Great Britain (Records Office) by communication 20th May 1985.
63. Ordnance Survey Map, scale 1:10,560, Hampshire Sheets XXII, Wiltshire LV, 1925 edn. revised 1923.
64. O'Hara, by communication 20th May 1985.
65. Maj. N.L. Chesher, G4 Quartering, HQ South-West District, by communication, 14th May 1985.

CHAPTER 9 BULFORD – THE SCENE UNFOLDS

1. The Methodist Church Forces Board, by communication, 31st October 1984.
2. Ordnance Survey map, scale 1:500, Wiltshire LV.9.17, 1903.
3. Methodist Forces Board, by communication.
4. 'Opening of the New Soldiers' Home at Bulford Camp' *Salisbury and Winchester*, 6th June 1903.
5. *Kelly's Directory for Wiltshire*, 1903, p.46.
6. Ordnance map, Wiltshire LV.9.17, 1903.
7. Ordnance Survey map, scale 1:2,500, Wiltshire LV.9, 1924.
8. *Thirteenth Report of the Church of England Board for the Welfare of Soldiers*, 1910, p.2.
9. *Fourteenth Report of Church of England Board*, 1911, p.2.
10. 'Young Mens' Christian Association', *Salisbury Journal*, 6th February 1904.
11. F. Stone, by communication, 4th May 1985.
12. Ibid.
13. Maj. J. Fairbank, Archivist, The Salvation Army, by communication, 9th April 1985.
14. 'Homes, Hostels and Hutments', *Under the Colours – The Salvation Army Magazine for Sailors and Soldiers*, November 1919.
15. 'Opening of the First Salvation Army Hut for the WAACs', *All the World* (A Salvation Army Magazine for the Forces), April 1918, p.151.

16. Firbank, by communication, 9th April 1985.
17. Miss H.M. Knox, General Secretary, Sandes Soldiers' and Airmen's Home, by communication 15th and 22nd May 1985.
18. N.E. Trowell, Hon. Curator, NAAFI Historical Collection, by communication, 6th November 1984.
19. Mrs. V.M. Edwards, Regional Secretary, NAAFI, Bulford, by communication, 21st June 1985.
20. Defence Land Agent, Durrington, by communication.
21. J.S. Gledhill, by communication.
22. Ibid.
23. Chief Education Officer, Wiltshire County Council, by communication, 4th July 1985.
24. Ibid.
25. Gledhill, by communication.
26. Ordnance map, Wiltshire, LV.9, 1924.
27. *The Monthly Army List*, October 1900, section (15).
28. *Army Lists*, January 1906–May 1907.
29. *Army Lists*, January 1904–October 1907.
30. Robin McNish, *Iron Division*, 1979, p.51.
31. Ibid., p.71.
32. Ibid., p.151.
33. Ibid., p.182.
34. Corporal R. Pointer, unpublished letters, October 1914.
35. Lieut. H.T.B. Drew, *The War Effort of New Zealand*, Wellington, 1923, p.244.
36. 'Bulford Between the Wars', *Bulford News*, November 1982, p.13.
37. C.E.H. Putt, New Zealand Expeditionary Force, drawing of Sling Camp Bulford, Salisbury Plain, England, 1914–1918 War.
38. Ordnance Survey Map, scale 1:10,560, Hampshire Sheet XXI. SW/Wiltshire Sheet LV.SW, 1926.
39. *Barrack Book 1929*, p.64.
40. Col. G.W.L. Nicholson, *Canadian Expeditionary Force 1914–1919*, Ottawa, 1964, p.32.
41. Ibid., p.36.
42. Ibid., p.39.
43. Col. H. Stewart, *The New Zealand Division 1916–1919*, Wellington, 1921, p.17.
44. Lieut. H.T.B. Drew, *The War Effort of New Zealand*, Wellington, 1923, pp.246–247.
45. Lieut.-Col. J. Studholme, *New Zealand Expeditionary Force: Record of Personal Services during the War*, Wellington, 1928, p.15.
46. Drew, *War Effort of New Zealand*, pp.248, 251.
47. Ibid., p.253.
48. Studholme, *Personal Services*, 1928, p.18.
49. Maj. W.J.A. Withers, 'Haere Mai, Haere Mai-Pakehas', *Bulford News*, February 1985, pp.14–15.
50. Ibid.
51. Col. P.W. Herring, Garrison Adjutant Bulford Camp, by communication.
52. Maj. J.R.B. Cook, 249 Signal Squadron (AMF(L)), by communication.
53. C.E.W. Bean, *Anzac to Amiens*, Canberra, 1948, p.518.
54. War Office Lands and Reconstruction Committee, *List of Lands and Buildings in the Occupation of the War Department*, 1st June 1918, – Wiltshire: Accommodation of Prisoners of War.
55. 'Bulford Between the Wars', *Bulford News*, November 1982, p.13.
56. Property Services Agency, Salisbury Plain Area Works Office, Bulford Village, by communication.
57. Maj. E. Clarke-Smith, *The Larkhill Guide*, Part I, *c.*1969, p.23.

58. Maj. N.L. Chesher, G4 Quartering, HQ South West District, by communication, 14th May 1985.
59. B. Bull, 'Bulford Camp – Did you know?', *Bulford News*, July/August 1984, p.6.
60. Army Order 78, 1938, 'Bulford-Re-naming of Barracks'.
61. Col. J.S. Fielding, by communication.
62. Ibid.
63. Mrs. J. Philpots, by communication.
64. Gavin Long, *To Benghazi* – Australia in the War of 1939–1945, Canberra, 1952, p.307.
65. Fielding, by communication.
66. Bull, *Bulford News*, July/August 1984, p.6.
67. N.D.G. James, *Gunners at Larkhill*, 1983, p.77.
68. T.H. Fitch, Custodian, Airborne Forces Museum, Aldershot, by communication, 19th July 1985.
69. Maj. Miles Whitelock, by communication, 5th May 1984.
70. Chesher, by communication.
71. Ordnance Survey Map, scale 1:10,560, sheet SV14SE, 1961.
72. Chesher, by communication.

CHAPTER 10 LARKHILL AND WEST DOWN

1. Lieut. A.W. Wilson, *The Story of the Gun*, 4th edn. 1976, p.47.
2. Ibid., p.49.
3. Ibid., p.50.
4. Ibid., p.53.
5. Ibid.
6. Major.-Gen. Sir Charles Callwell and Maj.-Gen. Sir John Headlam, *The History of the Royal Artillery*, vol.I, 1931, p.308.
7. *Annual Report of the School of Gunnery for Horse and Field Artillery at Home*, 1909, paras.11–15.
8. Callwell and Headlam, *History of the RA*, vol.I, 1931, p.229.
9. Ibid., p.233.
10. *Minutes of Proceedings of the War Office Salisbury Plain Committee*, p.63, minute 204.
11. Ibid., pp.106–107, minute 402.
12. Ordnance Survey Map: scale two inches to one mile, *War Department Land on Salisbury Plain*, 1903.
13. *Minutes of Salisbury Plain Committee*, p.122, minute 437.
14. *Annual Report of School of Gunnery*, 1901, para.87.
15. *Annual Report of School of Gunnery*, 1911, para.7.
16. Ordnance Survey Map: scale two inches to one mile, *War Department Land on Salisbury Plain*, published 1906, revised 1914.
17. E. Hall, by communication, 21st February 1984.
18. 'Salisbury Plain Notes', *The Salisbury and Winchester Journal*, 9th April 1904.
19. Ordnance Survey Map: *War Department Land on Salisbury Plain*, 1903.
20. Maj.-Gen. Sir John Headlam, *The History of the Royal Artillery*, vol.II, 1937, p.85.
21. 'Salisbury Plain Notes', *The Andover Advertiser*, 4th February 1910.
22. Ibid.
23. Ibid., 22nd April 1910.
24. Headlam, *History of the RA*, vol.II, p.31.
25. *Army Council Instruction 120*, 12th February 1915, 'Establishment of New School of Instruction for RH and RFA near Lark Hill, Salisbury Plain'.
26. Brig. F.E. Spencer, by communication, 5th April 1968.
27. *The Barrack Book 1929*, 118 General 714, The War Office 1929, pp.72, 74.
28. Brig. B.G. Mason, by communication, April 1968.
29. Maj. E. Clarke-Smith, *The Larkhill Guide*, *c.*1969, p.20.

30. A. Marshall, by communication, 18th May 1968.
31. War Office, Lands and Buildings Reconstruction Committee; *List of Lands and Buildings in the Occupation of the War Department*, 1st June 1918, Wiltshire: Schools, Rifle Ranges.
32. Ordnance Survey Map, scale 1:10,560 Wiltshire Sheet LIV. NW. 1926.
33. *The Barrack Book 1933*, 118 General 1331, 1933, p.74.
34. *The Barrack Book 1921*, p.68.
35. *The Barrack Book 1929*, p.74.
36. Ordnance Survey Map, scale 1:2,500, Wiltshire Sheet LIV.10, 1939.
37. War Office, *List of Lands and Buildings 1918*, Wiltshire: Accommodation of Prisoners of War.
38. Ibid., Schools.
39. N.D.G. James, *Gunners at Larkhill*, 1983, pp.48–50.
40. Col. G.W.L. Nicholson, *Canadian Expeditionary Force 1914–1919*, Ottawa, 1964, p.34.
41. Col. G.W.L. Nicholson, *The Fighting Newfoundlander 1700–1918*, 1964, pp.121–122.
42. Nicholson, *Canadian Expeditionary Force*, p.35.
43. Ibid., p.37.
44. Lieut. H.T.B. Drew, *The War Effort of New Zealand*, Wellington, 1923, p.244.
45. Nicholson, *Canadian Expeditionary Force*, p.36.
46. Nicholson, *Fighting Newfoundlanders*, p.125.
47. Nicholson, *Canadian Expeditionary Force*, p.36.
48. Nicholson, *Fighting Newfoundlanders*, p.124.
49. Nicholson, *Canadian Expeditionary Force*, p.36.
50. Ibid., p.34.
51. Ibid., p.39.
52. C.E.W. Bean, *The Australian Imperial Force in France 1916*, Sydney, 1929, p.168.
53. Ibid., p.169.
54. Ibid., p.176.
55. W.J. Sinney, unpublished letter, 3rd September 1916.
56. James Cowan, *The Maoris in the Great War*. Wellington, 1926, pp.158, 160.
57. Brig. B.G. Mason, personal account (unpublished) 1968.
58. *Army Council Instruction 660*, 4th December 1919, 'School of Artillery'.
59. 'Larkhill 1892–1976', *Plain Talk*, No.5, 1976, p.2.
60. Col. F.M. Hill, by communication, 18th September 1984.
61. Defence Land Agent, Durrington, by communication, 16th July 1984.
62. W.H. Russell, *Russell's Despatches from the Crimea 1854–1856* edited by Nicholas Bentley, 1966, p.135.
63. Ordnance Survey Map, scale 1:10,560, Wiltshire Sheet LIV.SE, 1926.
64. Brig. T.B. Trappes-Lomax, *Victoria County History – Wiltshire*, vol.III, 1956, p.97.
65. F. Stone, late Senior YMCA Representative Southern Command, *The Young Men's Christian Association with H.M. Forces on Salisbury Plain*, unpublished MS, October 1984.
66. Ibid.
67. 'Homes, Hostels and Hutments', *Under the Colours – The Salvation Army Magazine for Sailors and Soldiers*, November 1919.
68. N.E. Trowell, Hon. Curator, NAAFI Historical Collection, by communication, 6th November 1984.
69. Brig. C. Childs, 'One April Morning', *Gunner*, No.138, May 1982, pp.10, 11.
70. Brig. J.H.W.G. Richards, by communication, 5th July 1982.
71. K.E. Chant, by communication, 16th April 1985.
72. K.E. Chant, by communication, 25th April 1985.
73. Gavin Long, *To Benghazi* – Australia in the War of 1939–1945, Canberra, 1952, p.308.

74. W.G. McClymont, *To Greece*, Official History of New Zealand in the Second Word War 1939–1945, Wellington, 1959, p.33.
75. Col. C.P. Stacey, *Six Years of War – Official History of the Canadian Army in the Second World War,* Ottawa, 1955, vol.I, p.236.
76. Col. G.W.L. Nicholson, *More Fighting Newfoundlanders*, 1969, p.20.
77. Ibid., p.29.
78. Ibid., p.30.
79. Ibid., p.44.
80. Ibid., p.46.
81. Ibid., p.55.
82. Ibid., p.90.
83. Ibid., p.92.
84. Ibid., pp.110–111.
85. C. Fletcher-Holmes, G.G.P. Morgan and T. Smith by communication.
86. Col. J.A. Smith, Jnr. *Spearhead in the West*, pp.54, 56.
87. D. Kyte, Tilshead, by communication.
88. R. Kent, *First In – Parachute Pathfinder Company*, 1979, pp.1, 29.
89. Ordnance Survey Map, scale 1:10,560 sheet SU SW, 1961.
90. James, *Gunners at Larkhill*, pp.112–116.
91. 'Her Majesty The Queen's Visit to Larkhill', *Gunner*, No.31, June 1973, p.6.
92. 'HRH The Prince of Wales visits the Royal School of Artillery', *Gunner*, No.105, August 1979, p.3.
93. 'Views from Larkhill', *Gunner*, No.171, February 1985, p.11.
94. *Dictionary of National Biography*, vol.LII, 1897, pp.163–165.

CHAPTER 11 WARMINSTER, SUTTON VENY AND CODFORD

1. H. Graham, *The Annals of the Yeomanry Cavalry of Wiltshire*, 1886, p.3.
2. Ibid., p.5.
3. Ibid., p.7.
4. Ibid., p.13.
5. Ibid., p.14.
6. The Rev. J.J. Daniell, *The History of Warminster*, 1879, pp.143–144.
7. Graham, *Yeomanry Cavalry*, p.93.
8. Ibid., p.122.
9. Daniell, *History of Warminster*, p.147.
10. W. Richards. *His Majesty's Territorial Army*, *c.*1909, vol.IV, p.107.
11. Daniell, *History of Warminster*, p.147.
12. Richards, *HM Territorial Army*, vol.IV, p.108.
13. Daniell, *History of Warminster*, p.147.
14. Ordnance Survey Map: scale 1:2,500 Wiltshire Sheet LI4, 1941.
15. Ordnance Survey Map, GSGS 3906, scale 1:25,000, Sheet 41/16 SE, 1937.
16. Ordnance Survey Map: scale 1:2,500 Wiltshire Sheet LII, 1, 1887.
17. Ordnance Survey Map: scale 1:2,500 Wiltshire Sheet LII, 1, 1901.
18. Maj. F. Myatt, by communication, 7th December 1985.
19. 'New Warminster Barracks', *The Tank*, No.235, vol.20, November 1938, p.194.
20. 'The War – Miltiary Activity at Warminster', *The Warminster and Westbury Journal*, 7th August 1914.
21. *The Warminster and Westbury Journal*, 25th June 1915, p.5.
22. 'Regimental Sports at Warminster', *Warminster Journal*, 13th August 1915.
23. Brig. J.R.I. Platt, *The Royal Wiltshire Yeomanry (Prince of Wales Own)*, 1972, illustration no.2 following page 96.
24. Ordnance Survey Map GSGS 3906, scale 1:25,000, Sheet 41/16 SE, 1937.
25. Defence Land Agent, Durrington, by communication 15th June 1984.
26. 'New Warminster Barracks', *The Tank*, vol.I, 1938, pp.193–194.

27. 'Mr. Hore-Belisha opens New Barracks – Red Letter Day for Warminster', *The Warminster Journal*, 30th September 1938.
28. Ibid.
29. *The Tank*, No.235, vol.20, 1938, pp.193–194.
30. *The Tank*, No.245, vol.21, 1939, p.144.
31. Memorandum on the Royal Armoured Corps Training Regiments, September 1939 to May 1945 (undated), The Tank Museum Library, Bovington.
32. The Unit Diary, 27 Command Workshops.
33. D.C. Bateman, 'Incident at Imber', *After the Battle*, No.49, August 1985, pp.16–34.
34. Maj. J.J. Thornton, by communication, 19th November 1985.
35. *Spearhead in the West*, History of the 3rd Armored Division US Army, pp.54–55.
36. Sir Richard Hoare, Bt. *The Modern History of South Wiltshire*. vol.I, 1822, p.157.
37. Ibid.
38. Ibid.
39. *Kelly's Directory of Hampshire, Wiltshire, Dorsetshire, the Isle of Wight and the Channel Isles*, 1935, p.127.
40. Maj. A.A. Fairrie, 'Imber', *The Infantryman*, No.89, November, 1973, p.35.
41. Ibid., p.41.
42. Ibid.
43. Messrs. Woolley and Wallis' Catalogue for a sale held at Imber on 11th December 1943.
44. Messrs. John Jeffery & Son's Catalogue for a sale held at Edington, 15th December 1943.
45. Simon Appleyard and Bridget Cavanagh, 'Imber, Vanquished Village of Salisbury Plain', *This England*, Winter 1980, p.37.
46. Ibid.
47. 'War Office Skulduggery', *The Wiltshire Times*, 11th January 1985.
48. Maj. M.J. Fogwell, by communication, 7th January 1986.
49. Maj. F. Myatt, *The School of Infantry and Salisbury Plain*, 1981, p.25.
50. *The Monthly Army List*, September 1922.
51. Myatt, *School of Infantry*, pp.26–27.
52. Ibid., p.27.
53. Lieut.-Col. J.S. McLaren, by communication, 3rd December 1986.
54. Maj. R.B. Sullivan-Tailyour, by communication, 17th January 1986.
55. Property Services Agency, Warminster, by communication.
56.. Myatt, *School of Infantry*, pp.27–28.
57. Ibid., p.26.
58. *Defence Council Instruction 358/1969* – Staff College, Junior Division.
59. Property Services Agency, Warminster, by communication.
60. P.H. Vickers, Librarian Prince Consort's Army Library, by communication, 5th December 1985.
61. Maj. T.A. Gibson, 'Salisbury Plain and the Army', *The Infantryman*, No.77, October 1962, pp.40–41.
62. 'The Sutton Veny Camp – Erection of Huts', *Warminster Journal*, 6th November 1914.
63. Camps in the Codford Area taken from a Plan supplied by Lieut. Bruce RAMC. Clarence C. Hancock District Surveyor Warminster RDC, 20th May 1915. Wiltshire County Record Office.
64. *The War Graves of the British Empire – Wiltshire*, Imperial War Graves Commission, 1930, p.82.
65. Mrs. G. McCracken, *Looking Back on Seventy Years of Sutton Veny*, 1981, p.8.
66. Ibid., pp.8–9.
67. G. Ayres, Park View, Sutton Veny, by communication.
68. Ibid.

69. McCracken, pp.13–14.
70. 'Camp Notes', *Warminster Journal*, 26th July 1915.
71. 'Sutton Veny – More runaway trucks', *Warminster Journal*, 4th February 1916.
72. Maj.-Gen. Sir W.G. Macpherson, *Medical Services General History*, vol.I, 1921, p.78.
73. *War Graves – Wiltshire*, 1930, p.82.
74. Ibid.
75. Brig. J.R.I. Platt, by communication, 16th January 1986.
76. Ayres, by communication.
77. McCracken, p.23.
78. *Spearhead in the West*, pp.54, 62.
79. *Army and Air Force Installations outside Continental US*, 1947, p.170.
80. 'Lord Kitchener's Army – 24,000 coming to Codford', *Warminster Journal*, 11th September 1914.
81. 'The Codford Camps – Kitchener's Men under Canvas', *Warminster Journal*, 18th September 1914.
82. 'The Codford and Sherrington Camps', *Warminster Journal*, 25th September 1914.
83. 'The Wylye Valley Camps', *Warminster Journal*, 16th October 1914.
84. Camps in the Codford Area, Bruce and Hancock, 20th May 1915.
85. 'The Codford Camps – Railway Extensions in hand', *Warminster Journal*, 9th October 1914.
86. The Rev. E.D. Ginever, *The Ancient Wiltshire Village of Heytesbury*, revised by Dorothy Murray 1981, p.37.
87. *Warminster Journal*, 9th October 1914.
88. H. Cole, *The Village* 1914–1964, unpublished mss, *c.*1966, p.4.
89. *Warminster Journal*, 9th October 1914.
90. Cole, *The Village*, p.4.
91. Lieut. H.T.B. Drew, *The War Effort of New Zealand*, 1923, pp.246–247.
92. Ibid., p.123.
93. Macpherson, *Medical Services*, vol.I, p.78.
94. Drew, *War Effort of New Zealand*, p.266.
95. Ibid., pp.269–270.
96. Lieut.-Col. J. Studholme, *New Zealand Expeditionary Force: Record of Personal Services during the War*, Wellington, 1928, p.18.
97. *War Graves – Wiltshire*, 1930, p.77.
98. Ginever, *Village of Heytesbury*, p.21.
99. Michael Stratton, by communication.
100. Cole, *The Village*, p.3.
101. *Spearhead in the West*, p.54.
102. M. Stratton, by communication.

CHAPTER 12 WINGS OVER THE PLAIN

1. Brig. P.W.L. Broke-Smith, *The History of Early British Military Aeronautics*, 1952 (reprinted 1968), p.2.
2. Ibid., p.1.
3. Ibid., p.4.
4. Ibid., p.9.
5. Ibid., pp.11–12.
6. Ibid., p.22.
7. Ibid., p.21.
8. Ibid., p.30.
9. Ibid., p.29.
10. Ibid., p.31.

11. Ibid., p.32.
12. Ibid., p.33.
13. 'New Army Balloon', *Warminster Journal*, 8th May 1909.
14. N.D.G. James, *Gunners at Larkhill*, 1983, pp.21–22.
15. 'Shell Firing Experiments on Salisbury Plain', *Warminster Journal*, 8th May 1909.
16. Broke-Smith, *Early Aeronautics*, p.51.
17. N.C. Parker, by communication, 17th February 1982.
18. N.C. Parker, 'Army Airfields, No.3 – Larkhill', *The Army Air Corps Journal*, No.10, 1984, p.44.
19. Ordnance Survey Map: scale two inches to one mile, *War Department Land on Salisbury Plain*, 1906, War Department Property revised 1911.
20. Broke-Smith, *Early Aeronautics*, pp.40–41.
21. 'Aviation on Salisbury Plain', *Salisbury and Winchester Journal*, 4th February, 1911.
22. *History of Larkhill and Netheravon*, unpublished MS, undated, Air Historical Branch 5 (RAF).
23. *No.3 (Fighter) Squadron, RAF, A Short History*, unpublished MS, undated.
24. John W.R. Taylor, *C.F.S.*, 1958, pp.19–20.
25. No.3 Squadron, *A Short History*.
26. Taylor, *C.F.S.*, p.23.
27. Ibid., p.29.
28. RAF Personnel Management Centre, Officers Appointments and Records, 1B.
29. Peter Mead, *The Eye in the Air*, 1983, p.43.
30. N.C. Parker, 'Army Airfields, No.3,' p.49.
31. *History of Larkhill and Netheravon*.
32. Elizabeth Crittall, *Victoria County History – Wiltshire*, vol.XI, 1980, p.143.
33. 'Army Airman Killed at Upavon', *Warminster Journal*, 20th March 1914.
34. 'Two Army Airmen Killed', *Warminster Journal*, 14th August 1914.
35. 'Live Shell and Ball Cartridge – Realistic Exercises on Salisbury Plain – Scouting by Aeroplane', *Salisbury Journal*, 30th August 1913.
36. Capt. J.R. Cross, 'Army Airfields, No.2 – Netheravon', *The Army Air Corps Journal*, No.9, 1983, p.34.
37. Janet H. Stevenson, *Victoria County History – Wiltshire*, vol.XI, 1980, p.177.
38. *Army List*, January 1906.
39. Ibid., May 1907.
40. Ibid., October 1914.
41. Ibid., October 1919.
42. Ibid., September 1922.
43. Ibid., October 1922.
44. N.C. Parker, 'Army Airfields, No.3', p.49.
45. Cross, 'Army Airfields, No.2', pp.34–35.
46. T.H.J. Heffernan, *50 Years of Aeroplane and Armament Testing*, 1971, pp.5–6.
47. Cross, 'Army Airfields, No.2', p.35.
48. Heffernan, *50 Years*, p.7.
49. Brig. B.G. Mason, by communication, April 1968.
50. Heffernan, *50 Years*, p.7.
51. N.C. Parker, *Aviation in Wiltshire*, South Wiltshire Industrial Archaeology Society, Historical Monograph 5, 1982.
52. Capt. J.R. Cross, 'Royal Air Force Middle Wallop and the Army Air Corps Centre', *Army Air Corps Journal*, No.8, 1957–1982, p.40.
53. Christopher Ashworth, *Action Stations*, 5 Military airfields of the South-West, 1982, pp.142–145.
54. Ordnance Survey Map, scale 1:10,560 Wiltshire Sheet LIV. SW. 1926.
55. James, *Gunners at Larkhill*, p.43.
56. T.H.J. Heffernan, by communication, 6th February 1986.

57. Rt. Hon. D. Lloyd George, *War Memoirs of David Lloyd George*, vol.II, 1938, p.1098.
58. Ibid., p.1110.
59. Cross, 'Army Airfields, No.2, p.37.
60. Heffernan, *50 Years*, pp.8–9.
61. Vice Admiral Sir Arthur Hezlet, *Aircraft & Sea Power*, 1970, p.114.
62. Ibid., p.127.
63. Taylor, *C.F.S.*, p.110.
64. Ashworth, *Action Stations*, p.44.
65. Cross, 'Army Airfields, No.2', p.37.
66. *The Trechard Building Museum*, Programme for the Opening Ceremony, 9th September 1983.
67. Taylor, *C.F.S.*, p.149.
68. Cross, 'Army Airfields, No.2', p.37.
69. Heffernan, *50 Years*, p.13.
70. Ashworth, *Action Stations*, p.162.
71. War Office Lands and Reconstruction Committee, *List of Lands and Buildings in the Occupation of the War Department*, 1st June 1918, Wiltshire: Schools.
72. Ibid., Aerodromes.
73. G.H. Griffith, by communication, 25th April 1985.
74. K.E. Chant, by communication, 27th August 1985.
75. Griffith, by communication, 25th April 1985.
76. Ashworth, *Action Stations*, p.193.
77. H. Powey, by communication, 1st March 1968.
78. Lieut.-Col. J.B. Cowan, 'Observing Fire from Balloons', *Gunner*, No.166, September 1984, p.6.
79. Lieut.-Col. S.M.H. Nash, by communication, 10th September 1984.
80. Cowan, *Gunner*, September 1984, p.6. N.D.G. James, personal recollection.
81. 'A Short History of Royal Air Force Station, Middle Wallop', Museum of Army Flying.
82. Cross, 'RAF Middle Wallop', p.31.
83. 'Short History of RAF, Middle Wallop'.
84. Ashworth, *Action Stations*, pp.99–100.
85. Ibid., pp.81–82.
86. Ibid., p.139.
87. Ibid., pp.182–183.
88. Ibid., pp.162–163.
89. Ibid., pp.193–194.
90. Army Order 21, 27th February 1942, *Formation of the Army Air Corps*.
91. Col. P.R.R. de Burgh, by communication, 21st February 1986.
92. Ashworth, *Action Stations*, p.140.
93. Ibid., pp.191–192.
94. Heffernan, *50 Years*, p.15.
95. Cross, 'Army Airfields, No.2, p.37.
96. *Trenchard ... Museum*, Programme, 1983.
97. Cross, 'Army Airfields, No.2', p.37.
98. Taylor, *C.F.S.* pp.180–181.
99. Heffernan, *50 Years*, p.17.
100. Cross, 'RAF Middle Wallop', p.33.
101. Ibid., pp.33–34.
102. 'Testing to Perfection', *Air International*, vol.25, July 1983, p.17.
103. Taylor, *C.F.S.* p.190.
104. *Trenchard ... Museum*, Programme, 1983.
105. 'Short History of RAF, Middle Wallop'.
106. Army Order 82, 13th July 1957, *Army Air Corps – Formation*.

107. Cross, 'RAF Middle Wallop', p.37.
108. Heffernan, *50 Years*, p.30.
109. *Trenchard ... Museum*, Programme, 1983.
110. *Fiftieth Anniversary of Military Aviation*, Royal Air Force, Upavon, Saturday 16th June 1962, Programme of Events.
111. Cross 'Army Airfields, No.2', p.41.
112. Ibid., p.42.
113. Heffernan, *50 Years*, pp.26–27.
114. Cross, 'RAF Middle Wallop', p.40.
115. *Trenchard ... Museum*, Programme, 1983.
116. Upavon Site Plan, Air Ministry registered number 2513/37, 1937.
117. 'The Oldest Officers' Mess', *Royal Air Force Review*, September 1949.
118. Plan of the Central Flying School, Upavon. Scale 100 feet to an inch; issued by Northern Division Salisbury Plain District, 30th December 1917.
119. Ordnance Survey Map, scale 1:500, Wiltshire. Upavon RAF Camp Plan, 1925.
120. Ordnance Survey Map, scale 1:2,500, sheets SU 1454–1554, SU 1455–1555, SU 1655–1755, 1974.
121. Sqn. Ldr. K.W. Baldock, RAF, by communication.
122. Ordnance Survey Map, scale 1:10,560, Wiltshire Sheet XLVII, SE, 1925.
123. Maj. D.K.R. Clifton-Moore, by communication.
124. Cross, 'RAF Middle Wallop', p.39.
125. Clifton-Moore, by communication.
126. Ibid.

CHAPTER 13 OLD ROADS AND RAILWAYS OF THE PLAIN

1. Ordnance Survey Map to scale of one inch to one statue mile. Published 14th August 1817 by Col. Mudge at the Tower.
2. Ordnance Survey Map, scale 1:25,000, sheets SU05, SU15, 1958.
3. *A Narrative of the Life, Confession and Dying Speech of Jarvis Matchan who was on 2nd of August 1786, executed...*
4. Thomas Ingoldsby, 'The Dead Drummer', *The Ingoldsby Legends*, 4th edn., 1890, p.199.
5. Ordnance Survey Map, scale 1:25,000, sheets SU04, SU05, 1958.
6. *Annual Report of the School of Gunnery for Horse, Field and Heavy Artillery*, 1911, para.7.
7. *Annual Report of School of Gunnery, 1912*, Chief Instructor's letter.
8. Ordnance Survey Map, scale 1:25,000, sheets SU04, ST94, ST95, 1958.
9. Ordnance Survey Map, scale 1:25,000, sheet ST94, 1958.
10. J. Cary, 'A New Map of Wiltshire, 1801' – *New English Atlas*, 1809.
11. Brig. C. Childs, by communication, 30th December 1985.
12. Ernest F. Carter, *Railways in Wartime*, 1964, pp.16–18.
13. W.H. Russell, *Russell's Despatches from the Crimea 1854–1856* edited by Nicholas Bentley, 1966, p.160.
14. T.H.J. Heffernan, by communication.
15. *Minutes of Proceedings of the War Office Salisbury Plain Committee*, 1897–1902, p.7, minute 3.
16. Light Railway Commission, May, 1897, map showing the Pewsey & Salisbury Light Railway.
17. *Minutes of Salisbury Plain Committee*, p.9, minute 11.
18. Ibid., p.57, minute 183.
19. Ibid., p.64, minute 215.
20. Ibid., p.46, minute 131.
21. Ibid., p.122, minute 439.
22. Ibid., p.125, minute 446.

23. Ibid., pp.135–136, minutes 471, 472.
24. Ibid., p.25, minute 50.
25. Ibid., p.46, minute 130.
26. Ibid., p.57, minute 182.
27. Ibid., p.70, minute 238.
28. D. Bartholomew, *Midland & South Western Junction Railway*, vol.1, 1982, p.217.
29. T.B. Sands, *The Midland & South Western Junction Railway*, 1959 (reprinted 1979), p.18.
30. Bartholomew, *M & S W J R*, vol.1, p.219.
31. Sands, *M & S W J R*, p.18.
32. Bartholomew, *M & S W J R*, vol.1, pp.223–224.
33. Ibid., p.193.
34. Ordnance Survey Map, GSGS 2748, scale 1:20,000 published *c.*1918 and re-issued as GSGS 3906, scale 1:25,000 1931. Sheet 44/16 SE. Mapping and Charting Establishment RE, by communication October 1984.
35. Ordnance Survey Map, scale 1:10,560, Hampshire sheet XIV SE/Wiltshire sheet SLVIII SE, 1926 revised 1923.
36. J. Luke, Tidworth, by communication July 1984.
37. Ordnance Survey Map, scale 1:10,560, sheet SU24 NW, 1961, revised 1957.
38. Ordnance Survey Map, GSGS 3906, Sheets 44/16 SE, 44/14 NE.
39. Minutes of Salisbury Plain Committee, p.125, minute 446.
40. G.A. Pryer and A.V. Paul, *Track Layout Diagrams of the Southern Railway and B.R. (S.R.)*. Section S.3.
41. Ordnance Survey Map, scale 1:10,560, Wiltshire sheets LIV, SW, and LX, NW, 1926 revised 1923.
42. N.D.G. James, *Gunners at Larkhill*, 1983, pp.32–40.
43. 'The Wyley Valley Camps', *Warminster Journal*, 16th October 1914.
44. G. Ayres, Park View, Sutton Veny, by communication.
45. D. Burnett, *A Wiltshire Camera, 1914–1945*, 1976, illustration no.19.
46. *Warminster Journal*, 9th October 1914.
47. *Warminster Journal*, 16th October 1914.
48. Ordnance Survey Map, GSGS 3906, scale 1:25,000 Sheet 44/16 SW, 1931.
49. War Office Lands and Reconstruction Committee, *List of Lands and Buildings in the Occupation of the War Department*, 1st June 1918, Wiltshire; Light Railways and Sidings – Amesbury.
50. Ordnance Survey Map, GSGS 3906, Sheet 44/16 SE.
51. Ordnance Survey Map, scale 1:10,560, Wiltshire Sheet LXI.NW. 1926.
52. Ordnance Survey Map, GSGS 3906, Sheets 44/14 NW, 44/14 SW.
53. D. St. J. Thomas, *West Country Railway History*, vol.I, 4th edn. 1973, p.55.
54. Flt.-Lt. A.F. Saunders, 'The Little Trains of Chilmark', *The Railway Magazine*, vol.122, March 1976, pp.116–118.

CHAPTER 14 SALISBURY PLAIN TODAY

1. 'Salisbury Plain – All War Office Property to be sold', *The Wiltshire Gazette*, 10th April 1930.
2. 'Salisbury Plain', *The Wiltshire Gazette*, 17th April 1930.
3. 'Salisbury Plain – The Proposed War Office Purchases', *The Wiltshire Gazette*, 8th May 1930.
4. Commandant, Headquarters Salisbury Plain Training Areas, by communication, 7th July 1986.
5. Defence Land Agent, by communication, 7th July 1986.

6. *The War Graves of the British Empire – Cemeteries and Churchyards in the County of Wiltshire*, 1930. *The War Dead of the Commonwealth 1939–1945 – Cemeteries and Churchyards in the County of Wiltshire*, 1961.

Appendices

APPENDIX I. Acquisition of Land on Salisbury Plain 1897–1983

3rd Aug. 1897–31st Mar. 1902:	41,929 acres =	16,964 hectares.
1st Apr. 1902–31st Mar. 1910:	1,501 acres =	607 hectares.
1st Apr. 1910–31st Mar. 1920:	7,936 acres =	3,211 hectares.
1st Apr. 1920–31st Mar. 1930:	14,858 acres =	6,011 hectares.
1st Apr. 1930–31st Mar. 1940:	23,948 acres =	9,691 hectares.
1st Apr. 1940–31st Mar. 1950:	319 acres =	129 hectares.
1st Apr. 1950–31st Mar. 1960:	980 acres =	396 hectares.
1st Apr. 1960–31st Mar. 1970:	166 acres =	67 hectares.
1st Apr. 1970–31st Mar. 1983:	72 acres =	29 hectares.
	91,709 acres =	37,105 hectares

APPENDIX II. War Department and Defence Land Agents Salisbury Plain 1902–1985

1902–1909	T. Stirton
1909–1912	E. F. Slater
1912–1925	H. L. Pigott
1925–1938	A. P. Thorne
1938–1950	E. M. Mitchell
1950–1956	G. A. Gedge
1956–1960	E. R. Speak
1960–1966	C. L. Armitage
1966–1967	E. R. Speak
1967–1972	S.J. Cobb
1972–1974	D. L. Bowyer
1974–1975	G. V. Seymour
1975–1977	H. Cowan
1977–1985	C. Bayliffe
1985–	M.J. Mearns

APPENDIX III. Roads in Tidworth Garrison Bearing Commemorative Names

When the barracks and other buildings in Tidworth had been completed, the roads which served them were, with two exceptions, named after towns or villages in India and Afghanistan. These were arranged in alphabetical order from the west to the north-east of the barracks, the two exceptions being Bazaar Road and Grand Trunk Road. These roads are set out below with a brief note on the origin of their names.

Adampur Road
A village in the Punjab between Jullundur and Hoshiarpur, about 20 miles north-east of Jullundur.

Agra Road
A town in Uttar Pradesh (formerly the United Provinces) 110 miles south of Delhi and 150 miles west-north-west of Cawnpore (now Kanpur).

Baroda Road
Now known as Vadodara, Baroda lies 235 miles north of Bombay and 65 miles south-east of Ahmadabad.

Bazaar Road
In Indian towns, the bazaar was the equivalent of what is now commonly referred to as the shopping centre.

Cabul Road
The capital of Afghanistan which is also spelt Kabul is situated 120 miles west of the Khyber Pass and 235 miles north-east of Candahar.

Dasna Road
A small town 18 miles east of Delhi and 76 miles west-south-west of Moradabad in Uttar Pradesh.

Grand Trunk Road
The great Indian highway which ran for over 1500 miles from Calcutta to Peshawar, passing through Benares, Allahabad, Cawnpore, Delhi and Lahore. It was begun in 1839 and had been completely metalled by 1866 although several of the bridges carrying the road were not completed until later.

Jamrud Road
A village lying between Peshawar and the Khyber Pass, about 12 miles west of Peshawar.

Kirkee Road
A town about 5 miles to the north of Poona (now Pune). It was the scene of a battle between the British forces and the Mahratta prince, the Peshwa of Poona on 16th November 1817, during the Pindari war.

Khohat Road
A small town some 30 miles south of Peshawar and 95 miles west of Rawalpindi in Pakistan.

Lahore Road
A city in Pakistan situated 35 miles west of Amritsar and 50 miles north of Firozpur.

Lowa Road
An Indian village lying approximately midway between Lucknow and Rae Bareli which is itself almost halfway between Lucknow and Allahabad.

Margha Road
A village on the borders of Pakistan and Afghanistan about 40 miles west of Razmak and 85 miles south-east of Ghazni.

Meerut Road
A town in Uttar Pradesh, India, which is situated 38 miles north-east of Delhi. It was here that the Indian Mutiny began in 1857.

Nainital Road
A small town in the foothills of the Himalayas about 115 miles south-east of Dehra Dun and 70 miles north of Bareilly.

Nepaul Road
Nepaul which is also spelt Nepal, lies between India and Tibet, its southern boundary with India extending from a point some 60 miles east of Rampur to Darjeeling. Nepaul is the home of the Gurkhas many of whom have served with great distinction in both the Indian and British Armies.

Plassey Road
A village between Calcutta and Berampore about 25 miles south of the latter where Robert Clive (afterwards Lord Clive of Plassey) defeated the Nawab of Bengal on 23rd June 1757).

* * *

In recent years commemorative names have also been given to several other roads in Tidworth, a number of which are in the area of Sidbury Circular Road. These are given below.

Arcot Road
A small town about 70 miles west of Madras lying between Vellore and Kanchipuram which was captured by Clive when he defeated the Nawab of the Carnatic in 1751.

Jagdalik Road
A village in Afghanistan situated about 10 miles to the south of the road from Cabul to Jellalabad and about 35 miles east of Cabal.

Karachi Close
A city of Pakistan which stands to the north of the mouth of the River Indus and 90 miles south-east of Hyderabad.

Kohat Close
See Kohat Road.

Lahore Close
See Lahore Road

Peshawar Close
A town in Pakistan situated between Rawalpindi and the Khyber Pass some 35 miles east of the Pass.

Churchill Close
To commemorate Sir Winston Churchill, KG, PC, OM, CH. Prime Minister 1940–45 and 1951–55.

Paget Close
To commemorate Gen. Sir Bernard Paget, GCB, DSO, MC. C.-in-C. South Eastern Command, 1941, C.-in-C. Home Forces 1941–43, C.-in-C. 21st Army Group 1943, C.-in-C. Middle East Force 1944–46.

Wavell Close
To commemorate FM Earl Wavell, PC, GCB, GCSI, GCIE, CMG, MC., GOC.-in-c. Southern Command 1938–39, C.-in-C. Middle East 1939–41, C.-in-C. India 1941–43, Supreme Commander South West Pacific 1942, Viceroy and Governor-General of India 1943–47.

APPENDIX IV. Original Army Camps on Salisbury Plain

Camp	*Grid reference*	*Date of Ordnance Survey Map on which first recorded*	*Remarks*
Tidworth			
Tidworth Park	235469	1903	Revised 1899
Tidworth Pennings	224505	1903	Revised 1899
Park House	220448	1903	Revised 1899
Windmill Hill	251509	1903	Revised 1899
Perham Down	258491	1903	Revised 1899
Bulford			
Bulford Camp	190437	1903	Revised 1899
Brigmerston			
Bivouac Camp at Crossing 'C'	156459	1906	War Dept. property revised 1908.
Fittleton			
Bivouac Camp at Crossing 'A'	143500	1906	War Dept. property revised 1908.
Larkhill			
RFA Practice Camp	134444	1903	Revised 1899
Larkhill Camp	127443	1906	War Dept. property revised 1908.
Durrington Camp	134444	1906	War Dept. property revised 1908.
Fargo Down Camp	109442	1906	War Dept. property revised 1908.
Rollestone Camp	098453	1906	War Dept. property revised 1908.
Bustard Camp	091463	1906	War Dept. property revised 1908.
Hamilton Camp	108448	1906	War Dept. property revised 1914.
West Down			
Pond Farm	055534	1906	War Dept. property revised

Camp	*Grid reference*	*Date of Ordnance Survey Map on which first recorded*	*Remarks*
			1908.
West Down North	063494	1903	Revised 1899
West Down South	063480	1903	Revised 1899.

Note

a. The RFA Practice Camp at Larkhill was re-named Durrington Camp in 1906.

b. An Ordnance map (GSGS No. 3906 sheet 44/16 SE) issued in *c.* 1918, shows Durrington Camp astride the Packway, the grid reference of the centre of the camp area being 125444. In 1986 Roberts Barracks occupied a large part of this area.

c. It should be noted that the centre of the Camp shown on the 1918 map lies about 1,000 yds (914 m.) to the west of the area marked as Durrington Camp on the 1906 map.

d. These camps are shown on Maps 5 and 9.

APPENDIX V. Tidworth Tattoo 1935. Summary of the Programme

9.30 pm Fanfare by the Cavalry Trumpeters of the 2nd Cavalry Brigade.
- 4th/7th Dragoon Guards
- 3rd The King's Own Hussars
- 9th Queen's Royal Lancers
- 12th Royal Lancers

9.33 pm Beating Tattoo with the Massed Drums, Fifes and Bugles of the Southern Command.
- 2nd Bn. The King's Regiment
- 1st Bn. The Suffolk Regiment
- 1st Bn. The Green Howards
- 1st Bn. The Royal Welch Fusiliers
- 1st Bn. The Prince of Wales's Volunteers
- 2nd Bn. The Hampshire Regiment
- 1st Bn. The Royal Sussex Regiment
- 2nd Bn. The Loyal Regiment
- 1st Bn. The York and Lancashire Regiment
- 1st Bn. The Rifle Brigade.

9.40 pm Display by Motor Cycle Despatch Riders of the Royal Corps of Signals, Salisbury Plain Area.

9.50 pm Activity Ride by the 9th Queen's Royal Lancers.

9.56 pm First Entry of the Massed Bands of the Southern Command
- 4th/7th Dragoon Guards
- 3rd King's Own Hussars
- 9th Queen's Royal Lancers
- 12th Royal Lancers
- Royal Artillery (Salisbury Plain)
- 1st Bn. The Green Howards
- 2nd Bn. The King's Regiment
- 1st Bn. The Suffolk Regiment
- 1st Bn. The Royal Welch Fusiliers
- 1st Bn. The Prince of Wales's Volunteers
- 2nd Bn. The Hampshire Regiment
- 1st Bn. The Royal Sussex Regiment
- 2nd Bn. The Loyal Regiment
- 1st Bn. The York and Lancashire Regiment
- 1st Bn. The Argyll and Sutherland Highlanders
- 1st Bn. The Rifle Brigade
- The Royal Tank Corps.

10.11 pm Physical Training Display by 1st Bn. The Royal Sussex Regiment.

10.19 pm Musical Ride by the 4th/7th Dragoon Guards.

10.30 pm Infantry Drill by the 1st Bn. The Argyll and Sutherland Highlanders (Princess Louise's)

10.43 pm Display of Fireworks.

10.46 pm Second Entry of the Massed Bands.

11.01 pm Quebec, 13th September 1759 depicting the scaling of the Heights of Abraham, the battle with Montcalm and the death of Wolfe.

18th (Quebec 1759) Medium Battery, Royal Artillery
1st Bn. The Royal Welch Fusiliers
1st Bn. The Royal Sussex Regiment
2nd Bn. The Loyal Regiment
1st Bn. The Argyll and Sutherland Highlanders

11.16 pm Drive by 'L' (Néry) Battery, Royal Horse Artillery.

11.26 pm Grand Finale.

APPENDIX VI. Original Rifle Ranges on Salisbury Plain

Locality	*Range*	*Length in yards*	*Grid reference of firing points*
Bulford Camp	Bulford	A: 900 B: 1000 C: 1000 D: 900 E: 800	Between 203440 and 214451
Larkhill	Honey Down Barn	1000	077489
	Rushall Down	800	083504
	Shrewton Folly	A: 1000 B: 1000 C: 1000	In the area of 085474
	Wilsford Down	A: 1000 B: 1000	068524
Tidworth	Sidbury Hill	800	225497
Perham Down	Warren Hill	A: 1000 B: 1000	261488

Note

The above information has been obtained from an Ordinance Survey map entitled 'War Department Land on Salisbury Plain' which was published in 1903 and bears the endorsement 'Rifle ranges inserted 1900'.

APPENDIX VII. Lines and Barracks at Bulford Camp 1902–1985

1902 (a)	1922 (b)	1938 (c)	1985 (d)
A Lines	Mons Lines	Wing Barracks	Wing Barracks
B Lines	Cambrai Lines	Marlborough Barracks	Wing Barracks
C Lines	Arras Lines	Marlborough Barracks	Part Ward Barracks Part Gordon Barracks
D Lines	Ypres Lines	Beacon Barracks	Picton Barracks
E Lines	Aisne Lines	Gordon Barracks	Part Gordon Barracks Part Kiwi Barracks
F Lines	Messines Lines	Beacon Barracks	Picton Barracks
G Lines	Bapaume Lines	Gordon Barracks	Kiwi Barracks
ASC Lines	Amiens Lines	Ward Barracks	Ward Barracks
	Somme Lines	Ward Barracks	Ward Barracks
	Marne Lines	Kiwi Barracks	Kiwi Barracks
	Gallipoli Lines	Sling	Sling Barracks
	Baghdad Lines	Sling	Sling Barracks

Note with reference to columns (a) to (d)

(a) These are the designations of the original lines which are also shown in *The Barrack Book 1921*.

(b) The names inscribed on the Ordnance Map of 1924 (revised 1923) and in view of the entry in *The Barrack Book 1921*, they are assumed to date from 1922.

(c) The names authorised by Army Order 78 of 1938 'Bulford – Re-naming of Barracks' which also stated that the name 'Sling' was to denote an area and not a barracks. By 1961 Beacon had been re-named Picton and Marlborough had been divided between Ward and Gordon.

(d) Names in use in 1985 arising from the revision of barrack boundaries. Sling had now become 'Sling Barracks'.
These lines are shown on Map 7.

APPENDIX VIII. Barracks and Roads in Bulford Garrison Bearing Commemorative Names

The commemorative names which have been given to roads in the Garrison can be divided broadly into two categories. The major roads bear the names of certain distinguished officers of the First World War period, while those allocated to most of the secondary roads either refer to campaigns and battles of the First War or to towns which at the time were in the Commonwealth.

BARRACKS

Gordon Barracks

Named after Maj.-Gen. C. G. Gordon, CB who as a captain in the Royal Engineers, was attached to the Chinese Government and defeated the Taiping rebels in 1863–4. In January 1884 he was sent to evacuate the Egyptian forces from the Sudan but shortly after reaching Khartoum he was cut off by the supporters of the Mahdi. After a siege lasting 317 days the defences were overcome and Gordon was killed on 26 January 1885, two days before the relief column arrived.

Kiwi Barracks

To commemorate the New Zealand Forces that were stationed in Sling Camp during the First World War and on account of the kiwi which they cut out on Beacon Hill above the camp.

Picton Barracks

Picton Barracks, previously known as Beacon Barracks, takes its name from Lieut.-Gen. Sir Thomas Picton KCB who commanded the 3rd Division with considerable success during the Peninsular War (1808–13). He subsequently fought and was killed at Waterloo on 18 June 1815.

Sling Barracks

This barracks is named after the area in which it is partly situated and although this is not strictly a commemorative name, it is included so as to provide a complete list of the barracks.

Ward Barracks

Sir Edward Ward, Bt. KCVO, GBE, CB after whom this barracks is named, was a military administrator and soldier. After serving in the commissariat branch of the Government's Control Department, he was sent to the Sudan in 1885 as a civilian in connection with supplies, but was commissioned as a major into the Army Service Corps when it was formed in 1888. He was promoted to lieutenant-colonel and during the siege of Ladysmith, was responsible for feeding both troops and civilians and was described as 'the best supply officer since Moses'. Later he became Director of Supplies to Lord Roberts. His *Handbook of Army Service Corps Duties in Peace and War*, was published in 1897.

Wing Barracks

Wing Barracks and Wing Road commemorate Maj.-Gen. F. D. V. Wing CB who, in 1913, was CRA of the 3rd Division. After the battle of Le Cateau on 26th August 1914 he was given command of the 12th Division but was killed on 2 October 1915 near Loos being one of the first Divisional Commanders to become a casualty.

MAJOR ROADS

Haig Road

FM Earl Haig, KT, OM, GCB, KCIE, GCVO C-in-C Expeditionary Forces, France and Flanders, 1915–19.

Harrington Drive

Gen. Sir Charles Harrington GCB, GBE, DSO Deputy Chief of the Imperial General Staff, 1918–20; GOC in C the Army of the Black Sea, 1920–21; Governor and C-in-C Gibraltar 1933–38. Although not a major road, Harrington Drive has been included in view of its name.

Horne Road

Gen. Lord Horne, GCB, KCMG. Commanded 1st Army in France 1916; GOC in C Eastern Command 1919–23; Master Gunner of St. James's Park 1926–9.

Hubert Hamilton Road

Maj.-Gen. Hubert I. W. Hamilton, CB, CVO, DSO. Military Secretary to Lord Kitchener in South Africa (1900–2) and in India (1902–5). Commanded North Midland Division 1911–14.

Milne Road

FM Lord Milne, GCB, GCMG, DSO, Commanded 16th Army Corps in France 1915; Army of the Black Sea 1920; GOC-in-C Eastern Command 1923–6; CIGS 1926–33; Master Gunner of St. James's Park 1929–46.

Plumer Road

FM Viscount Plumer, GCB, GCMG, GCVO, GBE. Commanded 2nd Army, British Expeditionary Force in France 1915–17 and 1918; GOC Italian Expeditionary Force 1917; Governor and C-in-C Malta, 1919–24. Plumer Road was renamed Wyvern Road on 31 July 1977.

Rawlinson Road

Gen. Lord Rawlinson, GCB, GCSI, GCVO, KCMG. Commanded 4th Army British Expeditionary Force in France 1918; commanded forces in North Russia 1919.

Wing Road

This was originally named Grierson Road after Lieut.-Gen. Sir James Grierson, KCB, CMG, CVO who, after serving in Egypt, the Sudan, on the North-West Frontier and in South Africa commanded 1st Division, Aldershot

Command. He died in 1914. Information on Maj.-Gen. F. D. V. Wing, CB will be found under Wing Barracks.

SECONDARY ROADS

Baghdad Road

Baghdad was the chief city of Mesopotamia which now forms part of Iraq and of which Baghdad is now the capital. Failure to capture it in October 1915 led to the British Forces being beseiged at Kut al Amara which fell to the Turks on 29 April 1916. It was recaptured by Lieut.-Gen. Sir Stanley Maude, KCB, CMG, DSO on 24 February 1917 who then advanced on Baghdad which his troops entered on 11 March.

Gallipoli Road

On 25 April 1915, after preparatory action by the British and French navies, landings were made by Australian, British, French and New Zealand forces at several points on the Gallipoli Peninsula. Despite repeated attempts to overcome the Turkish defences the campaign proved to be a failure and on 8 January 1916 the last of the Allied Troops were evacuated.

Gaza Road

On 26 March 1917 British forces attacked the Turks who were holding this small town which is situated on the shores of the Mediterranean and about fifty miles south-west of Jerusalem. This did not succeed and another attempt, made three weeks later also failed. Finally on 2 November Gen. Sir Edmund Allenby (afterwards FM Viscount Allenby, GCB, CCMG, GCVO) who had taken over command, launched a new attack and the town was captured on 8 November 1917.

Marne Road

During the First World War two battles of the River Marne were fought. The first which took place on 8 September 1914 marked the end of the retreat from Mons and the mounting of an attack on the German forces which drove them back beyond the River Aisne. The second battle, on 20 July 1918, formed part of the Allied counter offensive in the summer of that year.

Mons Road

The battle of Mons which took place on 23 August 1914 was the first occasion when the British Expeditionary Force came into action against the German army, during the First War. On the following day the retreat from Mons began but three days later it was checked at Le Cateau although the situation was not finally restored until the first battle of the Marne.

Suvla Lane

On 6 August 1915, British troops landed at Suvla Bay which lies on the west coast on the Gallipoli Peninsula and about 30 miles north of Cape Helles.

Vimy Crescent

Vimy Ridge which lies between Lens and Arras was of great tactical importance to both sides in the First World War. It was finally captured by four Divisions of the Canadian Corps and the 17th British Corps during the battle of Arras which lasted from 9 to 14 April 1917.

APPENDIX IX. Service Museums on or near Salisbury Plain

The Museum of Army Flying
Middle Wallop, Stockbridge, Hampshire SO20 8DY
telephone: andover (0264) 62121 extn 421
Open: Daily 1000–1630 hrs (closed Christmas Day and New Year's Day)
Entrance on Andover–Salisbury road (A343) south of the main entrance to the Army Air Corps Centre.

The Tank Museum
Bovington Camp, Wareham, Dorset BH20 6JG
Telephone: Bindon Abbey (0929) 462721 extn 463
Open: Daily 1000–1700 hrs (closed for ten days at Christmas and the New Year)
Entrance in Bovington Camp. The location of the museum is indicated by directional signs.

The Trenchard Museum
Royal Air Force, Upavon, Pewsey, Wiltshire SN9 6BE
Telephone: Stonehenge (0980) 630351 extn 227
Open: By appointment. Intending visitors should telephone and write beforehand
Entrance in the Royal Air Force Station. Visitors are directed from the Guard Room.

The Museum of The Duke of Edinburgh's Royal Regiment
The Wardrobe, 58 The Close, Salisbury, Wiltshire SP1 2EX
Telephone: Salisbury (0722) 336222 extn 2683
Open: April–October: Sunday to Friday: 1000 to 1630 hrs
July–August: Saturdays as well: 1000 to 1630 hrs
November–March: Monday to Friday: 1000 to 1630 hrs
The museum occupies a most attractive house which dates from 1254 AD.

The Weapons Museum
The School of Infantry, Warminster, Wiltshire BA12 0DJ
Telephone: Warminster (0985) 214000 extn 2487
Entrance in the Garrison area. Visitors are directed from the entrance control point
Open: By appointment. Intending visitors should telephone or write to the Curator beforehand.
Open times:
Monday to Saturday 0900 to 1230 hrs
1400 to 1630 hrs.

APPENDIX X. Abbreviations

The following abbreviations which do not include decorations and ranks, have been used in some chapters.

AFV	Armoured Fighting Vehicle
AMF(L)	Allied Commander-in-Chief Europe Mobile Force (LAND)
ANZAC	Australian and New Zealand Army Corps
AOD	Army Ordnance Depot
ASC	Army Service Corps
ATS	Auxiliary Territorial Service
Bn	Battalion
CF	Chaplain to the Forces
ENSA	Entertainments National Services Association
FSMO	Field Service Marching Order
ft.	feet
GS	General Service
KRRC	King's Royal Rifle Corps
m.	metres
MT	Mechanical Transport
NAAFI	Navy, Army and Air Force Institutes
pr.	pounder
RA	Royal Artillery
RAC	Royal Armoured Corps
RAF	Royal Air Force
RAMC	Royal Army Medical Corps
RFA	Royal Field Artillery
SCF	Senior Chaplain to the Forces
SMLE	Short Magazine Lee-Enfield (rifle)
WAAC	Women's Auxiliary Army Corps
yds.	yards
YMCA	Young Men's Christian Association
YO	Young Officer
YWCA	Young Women's Christian Association

Index

Note. Aeronautical matters will be found under 'Flying'.